Health Econometrics
Using Stata

Health Econometrics
Using Stata

Partha Deb
Hunter College, CUNY and NBER

Edward C. Norton
University of Michigan and NBER

Willard G. Manning
University of Chicago

A Stata Press Publication
StataCorp LLC
College Station, Texas

Contents

Figures

Preface

This book grew out of our experience giving presentations about applied health econometrics at the International Health Economics Association and the American Society of Health Economists biennial conferences. In those preconference seminars, we tried to expose graduate students and early career academics to topics not generally covered in traditional econometrics courses but nonetheless are salient to most applied research on healthcare expenditures and use. Participants began to encourage us to turn our slides into a book.

In this book, we aim to provide a clear understanding of the most commonly used (and abused) econometric models for healthcare expenditure and use and of approaches to choose the most appropriate model. If you want intuition, meaningful examples, inspiration to improve your best practice, and enough math for rigor but not enough to cause rigor mortis, then keep reading. If you want a general econometrics textbook, then put down this book and go buy a general econometrics textbook. Get ready to try new methods and statistical tests in Stata as you read. Be prepared to think.

Despite years of training and practice in applied econometrics, we still learned a tremendous amount while working on this book from reading recent literature, comparing and testing models in Stata, and debating with each other. We particularly learned from our coauthor Will Manning, who unfortunately died in 2014 before seeing our collective effort come to fruition. Will was a fountain of knowledge. We think that his overarching approach to econometrics of repeated testing to find the best model for the particular research question and dataset is the best guide. The journey matters, not just the final parameter estimate.

In closing, we want to thank some of the many people who have helped us complete this book. David Drukker, editor and econometrician, had numerous suggestions, large and small, that dramatically improved the book. We are grateful to Stephanie White, Adam Crawley, and David Culwell at StataCorp for help with LaTeX, editorial assistance, and production of the book. We thank Betsy Querna Cliff, Morris Hamilton, Jun Li, and Eden Volkov for reading early drafts and providing critical feedback. We thank the many conference participants who were the early guinea pigs for our efforts at clarity and instruction and especially those who gave us the initial motivation to undertake this book. Our wives, Erika Bach and Carolyn Norton, provided support and encouragement, especially during periods of low marginal productivity. Erika Manning cared for Will during his illness and tolerated lengthy phone calls at odd hours, and Will's bad puns at all hours.

Partha Deb and Edward C. Norton

Notation and typography

In this book, we assume that you are somewhat familiar with Stata: you know how to input data, use previously created datasets, create new variables, run regressions, and the like.

We designed this book for you to learn by doing, so we expect you to read it while at a computer trying to use the sequences of commands contained in the book to replicate our results. In this way, you will be able to generalize these sequences to suit your own needs.

We use the `typewriter` font to refer to Stata commands, syntax, and variables. A "dot" prompt followed by a command indicates that you can type verbatim what is displayed after the dot (in context) to replicate the results in the book.

The data we use in this book are freely available for you to download, using a net-aware Stata, from the Stata Press website, http://www.stata-press.com. In fact, when we introduce new datasets, we load them into Stata the same way that you would. For example,

```
. use http://www.stata-press.com/data/heus/heus_mepssample
```

Try it. To download the datasets and do-files to your computer, type the following commands:

```
. net from http://www.stata-press.com/data/heus/
. net describe heus
. net get heus
```

The importance of the research and policy questions requires that we use the econometric models with care and that we think deeply about the correct interpretation. Faster computers do not obviate the need for thought.

In this book, we lay out the main statistical approaches and econometric models used to analyze healthcare expenditure and use data. We explain how to estimate and interpret the models using easy-to-follow examples. We include numerous references to the main theoretical and applied literature. We also discuss the strengths and weaknesses of the models we present. Knowing the limitations of models is as important as knowing when to appropriately use them. Most importantly, we demonstrate rigorous model testing methods. By following our approach, researchers can rigorously address research questions in health economics using a way that is tailored to their data.

1.1   Outline

This book is divided into three groups of chapters. The early chapters provide the background necessary to understand the rest of the book. Many empirical research questions aim to estimate treatment effects. Consequently, chapter 2 introduces the potential outcomes framework, which is useful for estimating and interpreting treatment effects. It also relates treatment effects to marginal and incremental effects in both linear and nonlinear models. Chapter 3 introduces the Medical Expenditure Panel Survey dataset, which is used throughout this book for illustrative examples. Chapter 4 illustrates how to estimate the average treatment effect, the treatment effect on the treated, and marginal and incremental effects for linear regression models. Chapter 4 also shows that misspecifications in OLS models can lead to inconsistent average effects. It also includes graphical and statistical tests for model specification to help decide between competing statistical models.

The core chapters describe the most prominent set of models used for healthcare expenditures and use, including those that explicitly deal with skewness, heteroskedasticity, log transformations, zeros, and count data. Chapter 5 presents GLMs as an alternative to OLS for modeling positive continuous outcomes. Generalized linear models are especially useful for skewed dependent variables and for heteroskedastic error terms. Although we argue that GLM provides a powerful set of models for health expenditure, we also lay out the popular log transformation model in chapter 6. Transforming a dependent variable by taking its natural logarithm is a widely used way to model skewed outcomes. Chapter 6 describes several versions that differ in their assumptions about heteroskedasticity and the distribution of the error term (normal or nonnormal). We show that interpretation can be complex, even though estimation is simple. Chapter 7 adds observations with outcomes equal to zero. Most health expenditure data have a substantial mass at zero, which makes models that explicitly account for zeros appealing. Here we describe and compare two-part and selection models. We explain the underlying assumptions behind the often misunderstood two-part model, and show how two-part models are superficially similar, yet strikingly different from selection models in fundamental ways. Chapter 8 moves away from continuous dependent variables to count models. These models are essential for outcomes that are nonnegative integer valued, including counts of office visits, number of cigarettes smoked, and prescription drug use.

The book then shifts to more advanced topics. Chapter 9 presents four flexible approaches to modeling treatment-effect heterogeneity. Quantile regression allows response heterogeneity by level of the dependent variable. We describe basic quantile regressions and how to use those models to obtain quantile treatment effects. Next, we describe finite mixture models. These models allow us to draw the sample from a finite number of subpopulations with different relationships between outcomes and predictors in each subpopulation. Thus, finite mixture models can uncover patterns in the data caused by heterogeneous types. Third, we describe local-linear regression, a nonparametric regression method. Nonparametric regression techniques make few assumptions about the functional form of the relationship between the outcome and the covariates and allow for very general relationships. Finally, conditional density estimation is another flexible alternative to linear models for dependent variables with unusual distributions. The last two chapters discuss issues that cut across all models. Chapter 10 introduces controlling for endogeneity or selection-on-unobservables of covariates of policy interest to the researcher. Chapter 11 discusses design effects. Many datasets have information collected with complex survey designs. Analyses of such data should account for stratified sampling, primary sampling units, and clustered data.

This book does not attempt to provide a comprehensive treatment of econometrics. For that, we refer readers to other sources (for example, Cameron and Trivedi [2005; 2010], Greene [2012], Wooldridge [2010; 2016]). Instead, we focus on healthcare econometric models that emphasize three core statistical issues of skewness, zeros, and heterogeneous response. We focus on providing intuition, a basic mathematical framework, and user-friendly Stata applications. We provide citations to the literature for original proofs and important applications. Much promising theoretical and applied work continues to appear in the literature each year. Jones (2010) describe some of the recent research as well as a wide range of econometric approaches.

1.2   Themes

Although we present numerous alternative models and ways to check and choose between those models, it should be no surprise that we do not determine a single best model for all situations or a good second-best model for all cases. Instead, researchers must find the model that is most appropriate for their research question and data. We recommend comprehensive model checking, but model checking is not a simple check list. It requires thought.

We aim to provide the tools to find the best model to consistently estimate the answer to the research question. This answer will often be a function of $E(y|\mathbf{x})$, such as the average treatment effect, or the marginal effect of a covariate on the outcome. We are also concerned about the precision of those estimates, measured by the variance of the estimators.

way that an experimental treatment is. In other analyses, the researcher may simply be interested in the best predictions of individual-level outcomes rather than the effect of a particular covariate on the outcome. In such analyses, the researcher would focus on prediction criteria to formulate an appropriate model. Such criteria may or may not be consistent with a model that is preferable in a causal analysis.

As we have suggested, researchers are often interested in the effects of a policy-modifiable treatment or intervention. Therefore, in this chapter, we provide a brief description of the potential-outcomes framework, beginning with the constructs in a completely general setting. We describe a number of ways to define different kinds of treatment effects. We also show how the framework can be cast in a regression setting, both linear and nonlinear in parameters. Finally, we describe in-sample specification testing and model-selection strategies, as well as out-of-sample cross-validation strategies to help guide choices of regression specification.

Estimating treatment effects in the potential-outcomes framework is an active area of research. This technical literature can be daunting, but there are also a number of excellent textbook descriptions that provide basic overviews, technical details, and examples (for example, in Wooldridge [2010] and Gelman and Hill [2007]). Imbens and Rubin (2015) devoted an entire book to this topic. There are also a number of surveys of this literature. The classic reference is Heckman and Robb (1985). Other, more recent surveys that take the reader closer to the frontiers of research in this area include Imbens (2004), Heckman and Vytlacil (2007), and Imbens and Wooldridge (2009).

In the following chapters, we describe a variety of linear and nonlinear regression models that work in many disparate situations for the estimation of treatment effects, for estimation of marginal effects, and for predictions of outcomes. Casting the problem at hand in a regression framework has many advantages, but it also opens numerous questions of exactly which regression model to choose in the final analysis. There are choices of what covariates to include and how to include them (polynomials, interactions), of how to specify the functional form of the conditional mean of the outcome (linear, log, or power), and of what statistical distribution to choose to complete the model.

In this chapter, we broadly describe the types of in-sample and out-of sample strategies a researcher might consider to answer such questions. We think it is important for researchers to have these strategies clearly laid out a priori so that they can make the best modeling choices in a systematic way. We describe details of model specification tests and model-selection methods in subsequent chapters. Cameron and Trivedi (2005) and Greene (2012) provide textbook descriptions of model selection and testing in nested and nonnested contexts. We also refer our readers to Claeskens and Hjort (2008) for detailed descriptions and comparisons of the vast literature on model selection both in- and out-of sample and to Rao and Wu (2001) and Kadane and Lazar (2004) for more technical descriptions and syntheses of the literature.

2.2 Potential outcomes and treatment effects

The potential-outcomes model of Rubin (1974) and Holland (1986) provides a framework to formally evaluate the conditions under which one can obtain a causal estimate of the effect of a binary treatment or intervention T on an outcome y. We describe this below, synthesizing the expositions of Wooldridge (2010) and Gelman and Hill (2007). Let the binary indicator T_i denote whether observation i received the treatment or not:

$$T_i = \begin{cases} 1 \text{ if observation } i \text{ received the treatment} \\ 0 \text{ if observation } i \text{ did not receive the treatment} \end{cases}$$

Following Rubin (1974), define potential outcomes y_i^0 and y_i^1 as the outcomes observed under control and treatment conditions, respectively. For an observation assigned to the treatment (that is, $T_i = 1$), y_i^1 is observed and y_i^0 is the unobserved counterfactual outcome, representing what would have happened to the unit if it had been assigned to the control condition. Conversely, for control observations, y_i^0 is observed, and y_i^1 is the counterfactual outcome. The potential outcomes may be continuous or discrete, nonnegative, positive, or real, etc. Indeed, we have specified nothing about the numeric and statistical properties of the potential outcomes. However, we do need to assume that treatment of observation i affects only the outcomes for observation i. This rules out spillover effects or externalities in the data-generating process of potential outcomes. In such a setting, the effect of treatment for observation i, denoted by τ_i, is defined as the difference between y_i^1 and y_i^0.

$$\tau_i = y_i^1 - y_i^0$$

The fundamental problem of causal inference is that we can generally observe only one of these two potential outcomes, y_i^0 and y_i^1, for each observation i. We cannot observe both what happens to an individual after being assigned to treatment (at a particular point in time) and what happens to that same individual after being assigned to the control condition (at the same point in time). In fact, we can relate the observed outcome (y_i) to the potential outcomes using the following relationship,

$$y_i = (1 - T_i)y_i^0 + T_i y_i^1 \tag{2.1}$$

which does not allow us to identify both of the two potential outcomes. Thus we can never measure a causal effect directly.

However, we can think of causal inference as a prediction of key features of the distribution of τ_i. The most commonly estimated feature is the average treatment effect (ATE), τ_{ATE}, calculated as

$$\tau_{\text{ATE}} = E\left(y_i^1 - y_i^0\right) = E\left(y_i^1\right) - E\left(y_i^0\right)$$

When the potential outcomes are also determined by other characteristics of the individuals (vector of covariates $\mathbf{x}_i$), the ATE conditional on $\mathbf{x}_i$, which is the vector of covariates for observation i, is simply

$$\tau_{\mathrm{ATE}}(\mathbf{x}_i) = E\left(y_i^1 - y_i^0 | \mathbf{x}_i\right)$$

Another commonly estimated effect is the average treatment effect on the treated (ATET), that is, the mean effect for those who were actually treated. This is equal to the ATE calculated only on the subsample of observations that received the treatment,

$$\tau_{\mathrm{ATET}} = E\left(y_i^1 - y_i^0 | T_i = 1\right)$$

The ATET can be extended to incorporate conditioning on $\mathbf{x}_i$:

$$\tau_{\mathrm{ATET}}(\mathbf{x}_i) = E\left(y_i^1 - y_i^0 | \mathbf{x}_i, T_i = 1\right)$$

How do we estimate these effects, given data on treatment assignment, observed outcomes, and covariates? The answer to this question depends on the design of the study, and—by implication—properties of the data-generating process that generates the potential outcomes. We describe estimating ATEs in three leading situations below: a laboratory experiment, a nonlaboratory experiment when randomization is possible, and an observational study without randomization.

2.3   Estimating ATEs

How might we observe or estimate the potential outcomes for an observation? In some situations, there might be close substitutes for the counterfactual outcomes. In other situations, it might be possible to randomly assign individuals to the treatment and control conditions so that the collection of control units could be viewed as a substitute for the counterfactual outcomes of the collection of treated units. In other situations, especially when close substitutes are unavailable and randomization is unfeasible, we may achieve similarity between treated and control units via statistical adjustment, for example, by linear or nonlinear regression.

Each of these approaches requires that an assumption of ignorability holds. In the language of basic statistics, ignorability means that the process by which observations are assigned to the treatment group is independent of the process by which potential outcomes are generated, conditional on observed covariates that partially determine treatment assignment and potential outcomes.

For more intuition, consider this example. Suppose that assignment to a healthcare checkup visit—the treatment group—is determined by the random outcome of a toss of a fair coin and also by an individual's age, gender, and whether he or she had a checkup last year. Suppose that the potential outcomes under the treatment and control conditions are also determined by the individual's age, gender, whether he or she had a checkup last year, whether he or she is in the treatment group, and a random error term.

Assignment to treatment would be considered ignorable conditional on age, gender, and an indicator for checkup last year because—conditional on those covariates—assignment to treatment is statistically independent of the potential outcomes. Now, imagine the same scenario in which we know age and gender but not whether the individual had a checkup last year. In this case, assignment to the treatment condition no longer satisfies ignorability, because one of the conditioning variables is not observed.

We maintain the ignorability assumption throughout most of the book. However, in chapter 10, we describe methods to obtain causal estimates when this assumption does not hold.

2.3.1 A laboratory experiment

In some situations, it may be possible to measure a close substitute for the counter-factual outcome. Such situations are quite common in the experimental sciences such as biology, chemistry, and physics, where bench scientists might subject a material to both treatment and control environments simultaneously, making the assumption that the samples of material subject to treatment and control are virtually the same in their response to the treatment and control conditions. They would observe y_i^0 from the sample unit subject to the control condition and y_i^1 from the virtually identical sample unit subject to the treatment condition. In this case, simple sample averaging of y_i^0 and y_i^1 would provide the estimates needed to construct ATEs.

Needless to say, such situations are more difficult to conceive in the context of human subjects. It might be possible to design an experiment in which one of a pair of identical twins is subjected to treatment and the other to the control condition as a way of obtaining a close substitute for the counterfactual outcome. However, such opportunities are rare.

2.3.2 Randomization

In a randomized trial, researchers begin with a pool of reasonably similar individuals, but not similar enough to allow the researcher to identify clone pairs as in the laboratory experiment described above. With such a reasonably homogeneous pool of individuals, a randomized design assigns those observations to the treatment and control conditions randomly. The treatment and control samples are not clone pairs, but they will have similar characteristics on average. In other words, when treatment is assigned completely at random, we can think of the different treatment groups (or the treatment and control groups) as a set of random samples from a common population. Then, because T_i is independent of y_i^0 and y_i^1, ATE and ATET are identical. To be precise, $\tau_{\text{ATE}} = \tau_{\text{ATET}}$ and $\tau_{\text{ATE}}(\mathbf{x}_i) = \tau_{\text{ATET}}(\mathbf{x}_i)$.

Randomization also provides a simple way to calculate estimates of expected potential outcomes, $E(y_i^0)$ and $E(y_i^1)$, using sample averages of expected observed outcomes, $E(y_i)$. To see this, consider that

$$E\left(y_i^1\right) = E\left(y_i^1 | T_i = 1\right) = E\left(y_i | T_i = 1\right)$$

where the fact that T_i is independent of y_i^1 is necessary to establish the first equality, and we use (2.1) to establish the relationship between potential and observed outcomes. Similarly,

$$E\left(y_i^0\right) = E\left(y_i^0 | T_i = 0\right) = E\left(y_i | T_i = 0\right)$$

If the treatment is randomized, we can obtain a consistent estimate of $E(y_i | T_i = 1)$ by calculating the mean of the observed outcome in the treated sample, and we can obtain a consistent estimate of $E(y_i | T_i = 0)$ by calculating the mean of the observed outcome in the control sample. We calculate the estimates of ATEs by applying these estimates of expected potential outcomes to the respective formula for the treatment effects of interest.

2.3.3 Covariate adjustment

When randomization is not possible, there will typically be self-selection into treatment. Individuals choose whether to receive treatment or not. In part their choice will be determined by the values of their observed covariates and, in part, by the values of their unobserved characteristics. For example, the decision to try a prescription drug (treatment) may depend on age (observable) and aversion to pain (unobservable).

As we described earlier in this section, ignorability is a key assumption required to obtain treatment effects in this context. Ignorability implies that the unobserved characteristics that determine selection into treatment are conditionally independent of the unobserved characteristics that determine the potential outcomes. Wooldridge (2010) describes these, and related conditions, in greater detail.

Denote $\mu_1(\mathbf{x}_i) = E(y_i^1 | \mathbf{x}_i)$ and $\mu_0(\mathbf{x}_i) = E(y_i^0 | \mathbf{x}_i)$. When conditional independence holds, it is still true that

$$\tau_{\mathrm{ATE}}(\mathbf{x}_i) = E\left(y_i^1 - y_i^0 | \mathbf{x}_i\right) = \mu_1(\mathbf{x}_i) - \mu_0(\mathbf{x}_i)$$

and

$$\tau_{\mathrm{ATET}}(\mathbf{x}_i) = E\left(y_i^1 - y_i^0 | \mathbf{x}_i, T_i = 1\right) = \mu_1(\mathbf{x}_i, T_i = 1) - \mu_0(\mathbf{x}_i, T_i = 1)$$

In addition, using (2.1) and the conditional independence assumption, we get $\mu_1(\mathbf{x}_i) = E(y_i | \mathbf{x}_i, T_i = 1)$ and $\mu_0(\mathbf{x}_i) = E(y_i | \mathbf{x}_i, T_i = 0)$. Therefore, as in the randomized trial case, we can estimate ATE and ATET using observed outcomes, observed treatment assignment, and observed covariates. It is possible to calculate estimates of $\mu_1(\mathbf{x}_i)$ and $\mu_0(\mathbf{x}_i)$ quite generally and even nonparametrically, but we will describe parametric regression-based methods below.

Before we do that, it is useful to see the general formula for estimates of $\tau_{\text{ATE}}(\mathbf{x}_i)$ and $\tau_{\text{ATET}}(\mathbf{x}_i)$, assuming one has consistent estimates of $\mu_1(\mathbf{x}_i)$ and $\mu_0(\mathbf{x}_i)$. In other words, $\tau_{\text{ATE}}(\mathbf{x}_i)$ is the sample average of the differences in estimated predicted outcomes in the treated and control states. The formula for the estimate of $\tau_{\text{ATE}}(\mathbf{x}_i)$ is

$$\widehat{\tau}_{\text{ATE}}(\mathbf{x}_i) = \frac{1}{N} \sum_{i=1}^{N} \{\widehat{\mu}_1(\mathbf{x}_i) - \widehat{\mu}_0(\mathbf{x}_i)\}$$

The formula for the estimate of $\tau_{\text{ATET}}(\mathbf{x}_i)$ in this case has a similar form, but this time, the sample over which we average is restricted to observations that received treatment. Formally,

$$\widehat{\tau}_{\text{ATET}}(\mathbf{x}_i) = \frac{1}{\sum_{i=1}^{N} T_i} \sum_{i=1}^{N} T_i \{\widehat{\mu}_1(\mathbf{x}_i) - \widehat{\mu}_0(\mathbf{x}_i)\}$$

2.4   Regression estimates of treatment effects

We now show how you can use regression models to estimate treatment effects. We remind our readers that if you are interested in estimating ATE or ATET, or those effects conditional on covariates, modeling efforts should focus on obtaining the best estimates of the conditional mean functions, $\mu_1(\mathbf{x}_i)$ and $\mu_0(\mathbf{x}_i)$. Consistency is clearly a desired property of the estimators, but precision is important as well. As is typical, there is often a tradeoff between consistency and efficiency of estimators, so we urge our readers to think through their modeling choices carefully before proceeding.

2.4.1   Linear regression

With the above general principles in mind, it is useful to begin with the randomization case even though no regression is necessary. In that case, we only need to estimate sample means. Nevertheless, we can also obtain an estimate of the ATE (which is equal to ATET) via a simple linear regression. Without any loss of generality, we can write the relationship between the observed outcome, y_i, and the treatment assignment, T_i, as

$$y_i = \beta_0 + \beta_T T_i + u_i$$

where β_0 and β_T are unknown parameters to be estimated and with $E(u_i) = 0$ for random error term u. In this case, $E(y_i^0) = \beta_0$ and $E(y_i^1) = \beta_0 + \beta_T$. So the ATE is β_T, and we can obtain its best linear unbiased estimate by an ordinary least-squares regression of y_i on T_i.

When observed covariates and unobserved characteristics determine selection into treatment and potential outcomes as we have shown above, the conditional independence assumption allows us to estimate ATEs using estimates of conditional means of observed outcomes. When we work with observational data, and even in large experiments carried out in natural environments, it is not always possible to achieve close similarity at the individual level or to be confident of randomization, especially if such an approach was pursued at group levels. In such situations, statistical modeling can help control for the effects of characteristics that are determinants of potential imbalances between treatment and control samples. For instance, by estimating a regression, we may be able to estimate what would have happened to the treated observations had they received the control, and vice versa, all else being held constant. Such an approach requires that the chosen regression specification be the correct data-generating process, or at least approximately correct.

The simplest, and perhaps most commonly used, linear (in parameters) regression specification is also specified as being linear in variables (the treatment indicator T_i and a vector of covariates $\mathbf{x}_i$):

$$y_i = \beta_0 + \beta_T T_i + \mathbf{x}_i' \boldsymbol{\beta}_{\mathbf{x}} + u_i \tag{2.2}$$

The ATE is $\tau_{\mathrm{ATE}}(\mathbf{x}) = \beta_T$, and its estimate is $\widehat{\tau}_{\mathrm{ATE}}(\mathbf{x}) = \widehat{\beta}_T$. Note that the inclusion of regressors in $\mathbf{x}$ that are either higher-order polynomial terms of covariates or interactions between covariates does not change the estimate of $\tau_{\mathrm{ATE}}(\mathbf{x}) = \beta_T$ solution in any essential way. Also, the estimate of the ATET is $\widehat{\tau}_{\mathrm{ATET}}(\mathbf{x}) = \widehat{\beta}_T$, which is the same as the ATE. However, unlike in the randomization case, this result is not a natural consequence of the study design. Instead, it is a restriction imposed by the functional form we chose for the regression. This functional form assumes that the treatment effect is identical for all; an alternative, shown below, interacts the treatment effect with covariates, allowing different treatment effects for different subpopulations.

If this regression specification adequately describes the data-generating process, we might comfortably conclude that ATE is equal to ATET. If not, we should enrich the specification. In chapter 4, we describe a number of specification checks and tests that help answer this question.

For now, consider a more general regression specification that relaxes the constraint of equality between ATE and ATET. Including terms in the regression specification that are interactions between $\mathbf{x}$ and T achieves this end. To see this, consider a model that includes a full set of interactions between covariates $\mathbf{x}$ and the indicator for treatment, T:

$$y_i = \beta_0 + \beta_T T_i + \mathbf{x}_i' \boldsymbol{\beta}_{\mathbf{x}} + (T_i \times \mathbf{x}_i') \boldsymbol{\beta}_{\mathbf{x}T} + u_i$$

In this specification, the expected outcome in the control condition ($T_i = 0$) is

$$\mu_i^0 = \beta_0 + \mathbf{x}_i' \boldsymbol{\beta}_{\mathbf{x}}$$

and the expected outcome in the treated condition ($T_i = 1$) is

$$\mu_i^1 = \beta_0 + \beta_T + \mathbf{x}_i' \boldsymbol{\beta}_{\mathbf{x}} + \mathbf{x}_i' \boldsymbol{\beta}_{\mathbf{x}T}$$

The difference between expected outcomes in the treated and control conditions is

$$\mu_i^1 - \mu_i^0 = \beta_T + \mathbf{x}_i' \boldsymbol{\beta}_{\mathbf{x}T}$$

Unlike the prior case shown in (2.2), the expected outcomes in treated and control cases and—consequently—the individual-level differences in expected outcomes are functions of the values of the individual's covariates, $\mathbf{x}_i$, leading to differences between ATE and ATET. Sample averages of estimates of individual-level differences in expected outcomes, over the entire sample for ATE and over the treated sample for ATET, are valuable. However, they may hide considerable amounts of useful information about how treatment effects vary across substantively interesting subgroups of the population. For example, the ATE of a checkup visit may be substantially different for men as opposed to women. Estimating two ATEs, one for the sample of men and the other for the sample of women, would provide a much richer understanding of the effect of this intervention than just one estimate.

This specification, a fully interacted regression model, raises concerns that the model is overspecified. A researcher may wonder whether there are too many extraneous variables in the specification. Perhaps only a few interactions are necessary. In the population, the coefficients on such extraneous variables would be zero. However, in finite samples, adding extraneous variables would decrease the precision of model estimates. In fact, estimates of ATEs from an overspecified model may be so imprecise that they render the point estimate relatively uninformative. Again, specification checks and tests described in chapter 4 could help answer this question.

The fully interacted regression specification is as general as you can make a linear-in-parameters regression specification for the conditional mean of the outcome. You can include as many covariates as you see fit and as many interactions between those covariates and polynomial functions of those covariates as you choose. Because the specification interacts every one of those terms with the binary indicator for treatment, it is akin to estimating two separate regressions with that specification of covariates— one for the sample of treated observations and the other for the sample of control observations.

However, these two model specifications are not quite equivalent. Estimating one fully interacted regression model on the entire sample assumes that the regression errors are homoskedastic. Estimating them separately allows the variance of the error terms to differ in the treated and control samples. This difference in the specifications of the variances of the errors does not change the point estimate of ATE or ATET. It will change the standard errors of those estimates, however; we will describe this in more detail and show examples in chapter 4.

2.4.2   Nonlinear regression

We can extend the regression approach to include statistical models that are nonlinear in parameters, such as most generalized linear models, logit, probit, Poisson, negative binomial, and other models for count data. We will describe a number of such models in detail in later chapters. For now, consider a nonlinear regression model in which

$$E\left(y_i|\mathbf{x}_i, T_i\right) = g(\beta_0 + \beta_T T_i + \mathbf{x}_i'\boldsymbol{\beta_x})$$

In this model, the covariates and the treatment indicator enter in a linear, additive way first, but then their effect on the outcome is transformed by a nonlinear function, $g(\cdot)$. In this setting, the individual-level expected treatment effect is no longer a linear function of covariates. Instead, it is

$$E\left(y_i^1|\mathbf{x}_i\right) - E\left(y_i^0|\mathbf{x}_i\right) = g\left(\beta_0 + \beta_T + \mathbf{x}_i'\boldsymbol{\beta_x}\right) - g\left(\beta_0 + \mathbf{x}_i'\boldsymbol{\beta_x}\right)$$

Once again, the individual-level expected treatment effect is a function of covariates $\mathbf{x}_i$, so it will vary from individual to individual across the sample. The estimation of the sample ATE is

$$\widehat{\tau}_{\text{ATE}}(\mathbf{x}_i) = \frac{1}{N}\sum_{i=1}^{N}\left\{g\left(\widehat{\beta}_0 + \widehat{\beta}_T + \mathbf{x}_i'\widehat{\boldsymbol{\beta}}_\mathbf{x}\right) - g\left(\widehat{\beta}_0 + \mathbf{x}_i'\widehat{\boldsymbol{\beta}}_\mathbf{x}\right)\right\}$$

where N denotes the sample size.

To estimate the ATET, we take the above formula but average only over the sample of treated observations. Here—as in nonlinear models generally—the individual-level expected treatment effect is a function of the covariates, $\mathbf{x}_i$, so expected treatment effects averaged over different samples will yield different estimates. Specifically, ATE will not be equal to ATET.

In each of the models described above, we have first described an ATE that averages the effects over the individuals in the sample and therefore over the distribution of covariates in the sample. We have also described the ATET, which requires averaging over the subsample of treated observations. In other cases, it may be insightful to calculate the ATEs for a hypothetical individual with a particular set of characteristics (covariates). For example, we may be less interested in comparing those with insurance with those without insurance over all individuals, and more interested in comparing those with insurance with those without insurance for individuals of lower socioeconomic status.

2.5   Incremental and marginal effects

So far, we have framed the researcher's problem as estimating the effect of a planned treatment or the effect of a treatment that naturally arises from a policy change. In both these cases, the treatment is a consequence of a planned intervention, and the policy question is whether that intervention had a desired effect. Nevertheless, many important empirical research questions are more descriptive in nature. For example, researchers may wish to know the difference in healthcare expenditures between men and women, or researchers may wish to know how many more doctor visits people take if they have some additional income, all else equal. Although these descriptive questions can also be framed as treatment effects, they are typically not described as such.

We will maintain an arguably artificial distinction between treatment effects of modifiable interventions and effects of other covariates that are not interventions and may not be modifiable. We use the phrase "incremental effect" to describe the effect of a change in an indicator variable—such as an individual's gender—and the phrase "marginal effect" to describe the effect of a small change in a continuous variable, such as an individual's income. Thus the average incremental effect of a binary indicator would be akin to the ATE; it would be calculated in exactly the same ways as described in section 2.4 in the contexts of linear and nonlinear regression models. The average marginal effect would also be akin to the ATE, but this time—instead of computing differences in outcomes—we would compute the derivative of the expected outcome with respect to the continuous covariate of interest.

To be more precise, let's first consider a linear regression model in which

$$E\left(y_i|x_i, d_i\right) = \beta_0 + \beta_x x_i + \beta_d d_i$$

where x_i is a continuous covariate and d_i is a binary indicator. The incremental effect of d_i is the discrete difference

$$E\left(y_i|x_i, d_i = 1\right) - E\left(y_i|x_i, d_i = 0\right) = \beta_d$$

The marginal effect of x_i is the derivative

$$\frac{\partial E\left(y_i|x_i, d_i\right)}{\partial x_i} = \beta_x$$

Both the average incremental effect and the average marginal effect are simply the coefficients on the respective variables in the regression. They are constant across the sample by definition.

Now consider a nonlinear regression model in which

$$E\left(y_i|x_i, d_i\right) = g(\beta_0 + \beta_x x_i + \beta_d d_i)$$

where x_i is a continuous covariate and d_i is a binary indicator covariate. The incremental effect of d_i is the discrete difference

$$E\left(y_i|x_i, d_i = 1\right) - E\left(y_i|x_i, d_i = 0\right) = g(\beta_0 + \beta_x x_i + \beta_d) - g(\beta_0 + \beta_x x_i) \tag{2.3}$$

The marginal effect of x_i is the derivative

$$\frac{\partial E\left(y_i|x_i, d_i\right)}{\partial x_i} = \beta_x \frac{\partial g(\beta_0 + \beta_x x_i + \beta_d d_i)}{\partial x_i}$$

Both the incremental effect and marginal effect will vary from individual to individual across the sample, because the function $g(\cdot)$ is nonlinear. We can calculate sample averages of these effects in a variety of ways, just as we can treatment effects.

Interaction terms see extensive use in nonlinear models, such as logit and probit models. Unfortunately, the intuition from linear regression models does not extend to nonlinear models. The marginal effect of a change in both interacted variables is not equal to the marginal effect of changing just the interaction term. More surprisingly, the sign may be different for different observations (Ai and Norton 2003). We cannot determine the statistical significance from the z statistic reported in the regression output. For more on the interpretation of interaction terms in nonlinear models and how to estimate the magnitude and statistical significance, see Ai and Norton (2003) and Norton, Wang, and Ai (2004).

In many of the examples we will use throughout this book, for simplicity, we will frame the underlying research question as being of a descriptive nature. Consequently, we will typically use incremental effects and marginal effects to describe the effects of interest. However, in each of those cases, especially if the researcher has a treatment in mind—but also in the purely descriptive situations—the formal potential-outcomes framework described here will provide invaluable insight into calculation and interpretation of effects.

2.6   Model selection

Once researchers have a good understanding of the parameters of interest, we recommend they examine the basic characteristics of the data they will use to estimate the parameters of a suitable econometric model. Some of the key questions involving the basic characteristics of the outcome of interest are as follows: Is the outcome always positive? Is it nonnegative with a substantial mass at zero? Is it integer valued? A second set of characteristics involves the nature of the statistical distribution of the outcome. Is the distribution of the outcome variable highly skewed? Is there good a priori reason to believe the parameter of interest varies across segments of the distribution of the outcome or on some other unobserved dimension? The answers to these questions will narrow down the class of models for consideration.

Next, the researcher should use a battery of specification checks and tests and model-selection criteria to narrow down the specification of the model along dimensions of specification of covariates, functional relationship between the outcome and covariates, and statistical distributions for the outcome (error). Needless to say, the researcher should revisit data characteristics and model choices if necessary. We view this as a critical component of a good empirical analysis, so we describe graphical checks and statistical tests throughout the book to help choose between alternative models. In this section, we provide an introduction to the model-selection approach we take throughout the book. We remind readers that there is a vast literature on this topic and refer them to Claeskens and Hjort (2008), Rao and Wu (2001), and Kadane and Lazar (2004) for further reading.

When the regression model is linear in parameters—or in the class of generalized linear models—the regression residuals form a basis for graphical checks of fit, which we demonstrate in chapter 4. Such checks can be extremely useful in detecting whether powers of covariates or interactions between covariates are necessary to specify the model correctly or whether a transformation of the outcome variable might improve the specification considerably.

However, although graphical tests are suggestive, they are not formal statistical tests. In chapter 4, we present three statistical tests for assessing model fit. The first two, Pregibon's (1981) link test and Ramsey's (1969) regression equation specification error test, directly test whether the specified linear regression shows evidence of needing higher-order powers of covariates or interactions of covariates for appropriate specification. The third—a modified version of the Hosmer–Lemeshow (1980) test—can be used generally, because it is based on a comparison between predicted outcomes from the model and model-free empirical analogs. If the model specification is not correct, then an alternative specification may predict better, indicating that the specification of the explanatory variables should change. When the modeling choices involve decisions such as adding covariates or powers and interactions of existing covariates, standard tests of individual or joint hypotheses (for example, the Wald and F tests) can also be useful.

2.6.1 In-sample model selection

The set of candidate models under consideration for an empirical application is often nonnested. This typically rules out standard statistical testing of model choice. In such situations, likelihood-based model-selection approaches are the most straightforward way to evaluate the performance of alternative models. Two model-selection criteria, which penalize the maximized log likelihood for the number of model parameters, are common: the Akaike information criterion (AIC) (Akaike 1970) and the Bayesian information criterion (BIC), also known as the Schwarz Bayesian criterion (Schwarz 1978). Both of these criteria have been shown to have many advantages in many circumstances, including robustness to model misspecification (Leroux 1992).

When the data have additional statistical issues, such as clustering and weighting, a strict likelihood interpretation of the optimand is often invalid. In most such situations, however, the model optimand has a quasilikelihood interpretation that is sufficient for these two popular model-selection criteria to be valid (Sin and White 1996; Kadane and Lazar 2004).

The AIC (Akaike 1970) is

$$\text{AIC} = -2\ln(L) + 2k$$

where $\ln(L)$ is the maximized log likelihood (or quasilikelihood) and k is the number of parameters in the model. Smaller values of AIC are preferable. The BIC (Schwarz 1978) is

$$\text{BIC} = -2\ln(L) + \ln(N)k$$

where N is the sample size. Smaller values of BIC are also preferable. For moderate to large sample sizes, the BIC places a premium on parsimony. Therefore, it will tend to

3.2   Overview of all variables

In this book's examples, we analyze various measures of annual expenditures and health service use. The example MEPS dataset has 44 variables and 19,386 observations on individuals who are age 18 years and older and have complete information on healthcare expenditures, counts of healthcare use, and key covariates. In addition, there are five variables that identify dwelling units, households and individuals in those households, and three variables that indicate features of the complex survey design that are useful when accounting for design effects (see chapter 11).

```
. use http://www.stata-press.com/data/heus/heus_mepssample
(Sample of MEPS 2004 data)
. *** 2004 MEPS data:  ID and sampling variables
. describe duid pid famidyr dupersid hieuidx wtdper var*

              storage   display    value
variable name   type    format     label     variable label
-------------------------------------------------------------------------------
duid            long    %12.0g               Dwelling unit id
pid             int     %8.0g                Person id (within dwelling unit)
famidyr         str1    %1s                  annual family identifier
dupersid        long    %12.0g               Person id (unique)
hieuidx         str7    %7s                  Health insurance eligibility unit
                                               id
wtdper          float   %9.0g                Sampling weight for person
varstr          int     %8.0g                Variance estimation for stratum
varpsu          byte    %8.0g                Variance estimation for psu
```

The demographic variables represent age, gender, race, ethnicity, family size, education, income, and region (see section 3.4 for further description and summary statistics).

```
. *** 2004 MEPS data:  Demographic variables
. describe age female race_* eth_hisp famsize ed_* lninc reg_*
              storage   display   value
variable name   type    format    label      variable label
```

variable name	storage type	display format	value label	variable label
age	byte	%8.0g		Age
female	byte	%9.0g	lb_female	
				Female
race_bl	byte	%14.0g	lb_race_bl	
				Black
race_oth	byte	%14.0g	lb_race_oth	
				Other race, non-white and non-black
eth_hisp	byte	%12.0g	lb_eth_hisp	
				Hispanic
famsize	byte	%8.0g		Size of responding annualized family
ed_hs	byte	%24.0g	lb_ed_hs	High school education
ed_hsplus	byte	%22.0g	lb_ed_hsplus	
				Some college education
ed_col	byte	%17.0g	lb_ed_col	
				College education
ed_colplus	byte	%8.0g	lb_ed_colplus	
				More than college education
lninc	float	%9.0g		ln(family income)
reg_midw	byte	%11.0g	lb_reg_midw	
				Midwest region
reg_south	byte	%9.0g	lb_reg_south	
				South region
reg_west	byte	%9.0g	lb_reg_west	
				West region

There are three health-status variables and four insurance variables (see section 3.4 for further description and summary statistics).

```
. *** 2004 MEPS data:  Health and insurance variables
. describe anylim mcs12 pcs12 ins_*
              storage   display   value
variable name   type    format    label      variable label
```

variable name	storage type	display format	value label	variable label
anylim	byte	%22.0g	lb_anylim	
				Any disability
mcs12	double	%10.0g		Mental health component of SF12
pcs12	double	%10.0g		Physical health component of SF12
ins_mcare	byte	%11.0g	lb_ins_mcare	
				Medicare insurance
ins_mcaid	byte	%11.0g	lb_ins_mcaid	
				Medicaid insurance
ins_unins	byte	%13.0g	lb_ins_unins	
				Uninsured
ins_dent	float	%19.0g	lb_ins_dent	
				Dental insurance, prorated

There are nine measures of expenditures and six measures of healthcare use. We often use these variables as dependent variables in the examples (see section 3.3 for further description and summary statistics).

```
. *** 2004 MEPS data:  Expenditures and use variables
. describe exp_* use_*

                storage   display   value
variable name   type      format    label     variable label
```

```
exp_tot         long      %12.0g              Total medical care expenses
exp_ip          float     %9.0g               Inpatient expenses = exp_ip_fac +
                                                 exp_ip_md
exp_ip_fac      long      %12.0g              Inpatient facility expenses
exp_ip_md       int       %8.0g               Inpatient md expenses
exp_er          int       %9.0g               ER expenses = exp_er_fac +
                                                 exp_er_md
exp_er_fac      int       %12.0g              ER facility expenses
exp_er_md       int       %8.0g               ER md expenses
exp_dent        int       %8.0g               Dental care expenses
exp_self        long      %12.0g              Total expenses paid by self or
                                                 family

use_disch       byte      %8.0g               # hospital discharges
use_los         int       %8.0g               # nights in hospital
use_er          byte      %8.0g               # emergency room visits
use_off         int       %8.0g               # office-based provider visits
use_dent        byte      %8.0g               # dental visits
use_rx          int       %8.0g               # prescriptions and refills
```

3.3   Expenditure and use variables

All the expenditure and use variables are highly skewed, with a large mass at zero. Expenditures include out-of-pocket payments and third-party payments from all sources. They do not include insurance premiums. We measure all expenditures in 2004 U.S. dollars; adjusting expenditures for inflation to 2016 would increase nominal amounts by about 27%. There are several ways to provide summary statistics for each type of expenditure. First, we provide summary statistics on all observations, including those with zeros. We report skewness and kurtosis to show how skewed these variables are. None of the summary statistics are corrected for differential sampling or clustering. Total annual expenditures on all healthcare averaged $3,685 (in 2004 dollars), with a range from $0 to $440,524. Inpatient expenditures averaged $1,123. Inpatient expenditures are divided into inpatient facility and inpatient physician expenses. The total amount paid by a family or individual was less than $700 on average but was as high as $50,000.

```
. *** Summary statistics for expenditure variables
. tabstat exp_tot exp_ip exp_ip_fac exp_ip_md exp_er exp_er_fac exp_er_md
>         exp_dent exp_self,
>         stats(mean sd skewness kurtosis min max) columns(statistics) varwidth(8)
```

variable	mean	sd	skewness	kurtosis	min	max
exp_tot	3685.25	9768.475	13.84523	414.4841	0	440524
exp_ip	1122.972	7283.09	22.12843	845.3283	0	376987
exp_ip~c	964.8604	6784.896	24.54683	999.7019	0	364292
exp_ip~d	158.112	868.0844	11.03482	184.4009	0	23203
exp_er	130.1588	685.5471	12.68341	238.7315	0	20545
exp_er~c	108.3229	624.8073	13.91178	283.1294	0	20468
exp_er~d	21.83596	110.8629	15.43297	566.776	0	6150
exp_dent	211.2738	657.1742	7.071237	79.83962	0	16275
exp_self	685.2889	1468.705	9.017003	179.3081	0	50850

Next, we construct dummy variables that equal one for observations with zero expenditures to help show the distributions of these variables. We give them names ending in 0. Summarizing them shows the fraction of the sample without any expenditure. Although the majority have some healthcare expenditures, more than 17% do not. The majority of individuals in the sample have no inpatient stays, emergency room visits, or dental expenditures.

```
. *** Percent of observations with zero expenditures
. tabstat exp_tot0 exp_ip0 exp_er0 exp_dent0 exp_self0,
>         stats(mean) columns(statistics) varwidth(14)
```

variable	mean
exp_tot0	.1774476
exp_ip0	.9096771
exp_er0	.862736
exp_dent0	.6265862
exp_self0	.2208295

Finally, we show summary statistics (including the coefficient of skewness) for the subset with positive values (different N for each variable), both for the raw variable and the logged variable. We give them names ending in gt0. The raw positive expenditure variables have extremely high skewness, with values ranging from 4.6 to almost 13.

```
. *** Summary statistics for expenditure variables, if any expenditures
. tabstat exp_totgt0 exp_ipgt0 exp_ergt0 exp_dentgt0 exp_selfgt0,
>         stats(n mean sd skewness min max) columns(statistics)
>         varwidth(13)
```

variable	N	mean	sd	skewness	min	max
exp_totgt0	15946	4480.262	10604.14	12.94196	2	440524
exp_ipgt0	1751	12432.86	21139.41	8.269471	20	376987
exp_ergt0	2661	948.2371	1627.563	5.276533	3	20545
exp_dentgt0	7239	565.79	977.7818	4.63159	3	16275
exp_selfgt0	15105	879.5108	1611.725	8.415801	1	50850

The logged expenditures have skewness much closer to zero, implying that the distribution is closer to being lognormal. However, this is not a formal test; we cover the Box–Cox test of normality in section 6.5.

```
. *** Summary statistics for logarithm of expenditure variables
. tabstat lnexp_tot lnexp_ip lnexp_er lnexp_dent lnexp_self,
>         stats(n mean sd skewness min max) columns(statistics)
>         varwidth(13)
    variable |        N       mean        sd   skewness        min        max
-------------+------------------------------------------------------------------
   lnexp_tot |    15946   7.234585   1.640616  -.2364216   .6931472   12.99572
    lnexp_ip |     1751   8.811378   1.138532  -.5018942   2.995732   12.83997
    lnexp_er |     2661   6.062692   1.305349  -.2472986   1.098612   9.930373
  lnexp_dent |     7239   5.560598   1.187932   .3775276   1.098612   9.697386
  lnexp_self |    15105   5.771392   1.595714   -.513215          0   10.83664
```

The distribution of the logarithm of positive values of total healthcare expenditures looks much more symmetric than the distribution of positive expenditures (see figure 3.1). Although it is tempting to conclude that the distribution of the logarithm of expenditures is normal, or truly symmetric, both of those conclusions are typically wrong; modeling expenditures as such can lead to incorrect conclusions. One of this book's main themes is to model such variables.

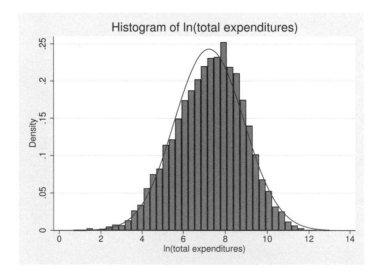

Figure 3.1. Empirical distribution of ln(total expenditures)

The example dataset has six variables that measure healthcare use (discharges, length of stay, three kinds of office-based visits, and prescriptions). Each use variable has a large mass at zero and a long right tail. On average, people had nearly 6 office-based provider visits, almost 13 prescriptions (or refills), and about 1 dental visit. About 29% have no office-based provider visits during the year, and 5% have at least 25. One-third have no prescriptions or refills during the year, while 17% have at least 25. Well over half the sample report having no dental visits during the past year. About 60% of adults have no dental visits, while closer to 30% have no office-based provider visits, prescriptions, or refills.

```
. *** Summary statistics for use variables
. tabstat use_*, stats(n mean sd skewness min max) columns(statistics)
    variable |        N        mean          sd  skewness         min        max
-------------+--------------------------------------------------------------------
   use_disch |    19386    .1219437    .4528718  5.784342           0          9
     use_los |    19386    .6429382    5.437346  32.25693           0        365
      use_er |    19386    .2144847    .6528512  5.915719           0         14
     use_off |    19386    5.802383    10.86976  5.549091           0        187
    use_dent |    19386    .9038997     1.68107  3.432859           0         22
      use_rx |    19386    12.81224    22.78304  3.480176           0        337
```

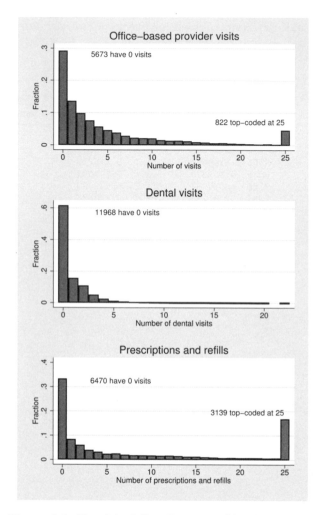

Figure 3.2. Empirical distributions of healthcare use

The density for all six use variables falls gradually for nearly the entire range. Histograms for three of the use variables (office-based visits, dental visits, and prescriptions and refills) show a large mass at zero and a declining density for positive values (see figure 3.2). We top coded some values at 25 for the purpose of the histogram.

3.4   Explanatory variables

The explanatory variables in the dataset include demographics, education, income, and geographic location. The average age is 45 and ranges from 18 to 85. However, AHRQ top-coded age at 85 because of confidentiality concerns. Just over half the sample is female. The vast majority is white (80%). About 14% are African American, and the remaining 6.5% comprise an other-race category. The race and ethnicity variables are not mutually exclusive. The distribution of race and ethnicity variables reflects oversampling of minority groups. Family size averages 3.0 but is as high as 13.

Years of education is coded as a sequence of categorical variables. About 30% have no more than a high school degree, and another 30% have at least a college degree. The natural logarithm of household income is fairly symmetric, with a mean of about 10.6. Household income averages nearly $60,000 and is quite skewed. At the high end are the 398 households with annual income greater than $200,000, and at the bottom end are the 498 households with annual income less than $6,000.

```
. *** Summary statistics for demographic variables
. summarize age female race_* eth_hisp famsize ed_* lninc reg_*

    Variable |       Obs        Mean    Std. Dev.        Min         Max
-------------+-----------------------------------------------------------
         age |    19,386    45.36088       17.387         18          85
      female |    19,386     .5495719     .4975494          0           1
     race_bl |    19,386     .1382441     .3451649          0           1
    race_oth |    19,386     .0653564     .2471601          0           1
    eth_hisp |    19,386     .2262973     .4184446          0           1
-------------+-----------------------------------------------------------
     famsize |    19,386     3.004488     1.659554          1          13
       ed_hs |    19,386      .297328     .4570939          0           1
    ed_hsplus|    19,386      .258795     .4379841          0           1
      ed_col |    19,386      .170587     .3761574          0           1
   ed_colplus|    19,386     .1272568      .333269          0           1
-------------+-----------------------------------------------------------
       lninc |    19,386    10.64533     .9260743   1.609438    12.98966
    reg_midw |    19,386     .1986485     .3989931          0           1
   reg_south |    19,386     .4096255     .4917773          0           1
    reg_west |    19,386     .2449706     .4300809          0           1
```

There are three health measure variables. One is a dichotomous measure of whether the person has any limitations, based on activities of daily living and instrumental activities of daily living. About 28% of the sample has at least 1 limitation. The other two health measures are based on the physical and mental health components of the Short Form 12. They are used to construct continuous measures on a scale from 0 to 100, with a mean of about 50. A higher number indicates better health. Both distributions are skewed left, with a median three to four points above the mean.

General health insurance is divided into four categories (with private insurance being the omitted group). About 19% are covered by Medicare, 14% by Medicaid, and 49% by private insurance. The remaining 18% are uninsured. There are 760 observations dually eligible for both Medicare and Medicaid. In addition to regular health insurance, 40% of the sample have prorated dental insurance, giving some observations fractional values.

```
. *** Summary statistics for health and insurance variables
. summarize anylim mcs12 pcs12 ins_*

    Variable |        Obs         Mean     Std. Dev.       Min         Max
-------------+--------------------------------------------------------------
      anylim |     19,386     .2845868      .4512291         0           1
       mcs12 |     19,386     50.22171      10.19464      1.35       75.06
       pcs12 |     19,386     49.01453      11.01185      4.56       72.17
   ins_mcare |     19,386     .1864232      .3894578         0           1
   ins_mcaid |     19,386     .1442278      .3513296         0           1
-------------+--------------------------------------------------------------
   ins_unins |     19,386     .1776024      .3821875         0           1
    ins_dent |     19,386     .3967468      .4684185         0           1
```

3.5   Sample dataset

Interested readers can use the example dataset based on the 2004 MEPS data to reproduce results found in this book. The sample from the 2004 full-year consolidated data file includes all adults ages 18 and older and who have no missing data on the main variables of interest. There are 19,386 observations on 44 variables. This dataset is publicly available at http://www.stata-press.com/data/eheu/dmn_mepssample.dta.

As stated in the introduction, we created the example dataset in 2008 for illustrative purposes only. It is not intended for research and does not include any updates that AHRQ made to the file since that time. The dataset does not include poststratification weights to reflect sample loss due to partial years of participation or item nonresponse. Interested readers should use this sample dataset for learning purposes only and should obtain the most recent version of MEPS to conduct any research.

3.6   Stata resources

Stata has excellent documentation for users to learn commands. To get started, see the *Getting Started With Stata* manual—which introduces the basic commands and interface—and the *Stata User's Guide*, which has a brief overview of the essential elements of Stata and practical advice. At the end of each remaining chapter, we highlight the commands used in the chapter and where to find more information about them in the Stata manuals.

Use the `describe` command to describe each variable's basic characteristics, including the often informative label. The command for basic summary statistics is `summarize`; use `tabstat` or `summarize` with the `detail` option to generate more extensive statistics. Two of the most commonly used commands for data cleaning are `generate` and `replace`. See the *Data-Management Reference Manual* for commands to describe, generate, and manipulate variables.

To visually inspect variable distributions, use the `histogram` command. To create scatterplots of variables that show their relationship visually, use the `scatter` command. See Stata's *Graphics Reference Manual* for all graphing commands.

4.3.3 Treatment effects

Next, we interpret the dichotomous variable indicating if the person has a
For purposes of illustration, we will consider `anylim` to be a treatment
treatment variable, the typical goal is to estimate the average treatmen
and the average treatment effect on the treated (ATET) (see chapter 2)
do this in Stata is to estimate an incremental effect using the `margins`
the `contrast()` option. Another way is to use the Stata treatment-effe
`teffects`. We will demonstrate both ways to clarify how these similar
estimate the same magnitude of treatment effects, and we will explain wh
standard errors are slightly different.

First, we use the results from the OLS regression model to estimate pr
comparing predictions that everyone had a limitation with predictions tl
limitations. By this approach, we see that the average predicted spe
one had any limitations is only $3,030, while predicted spending as if
limitation is $7,487.

```
. *** Margins for any limitations with sample-average standard error
. margins i.anylim

Predictive margins                              Number of obs       =
Model VCE      : Robust

Expression     : Linear prediction, predict()
```

	Margin	Delta-method Std. Err.	t	P>|t|	[95% Conf.
anylim					
No activi..	3030.488	61.61169	49.19	0.000	2909.722
Activity ..	7487.416	224.2832	33.38	0.000	7047.796

We use the `contrast()` option to take the difference between those
margins. When everyone rather than no one, has limitations, average e
crease by $4,457 = $7,487 − $3,030; this is exactly equal to the OLS estim:
The delta-method standard errors take the covariates as fixed, which c
sample ATE (Dowd, Greene, and Norton 2014). Later in this section, we
to compute standard errors for a population-averaged treatment effect,
for the fact that the covariates also have sampling variation in the popu

4 The linear regression model: Specification and checks

4.1 Introduction

The linear regression model is undoubtedly the workhorse of empirical research. Researchers use it ubiquitously for continuous outcomes and often for count and binary
outcomes. With relatively few assumptions—namely, the relationship between the outcome and the regressors is correctly specified, and the error term has an expected value
of zero conditional on the values of the regressors—ordinary least-squares (OLS) estimates of the parameters of the model are unbiased and consistent. In other words, given
those two assumptions, OLS delivers estimates that are correct on average. In addition,
if the distribution of the errors have constant variance across the sample observations
and are uncorrelated across sample observations, then OLS produces estimates that have
the smallest variance among all linear unbiased estimators.

The formal statement of these properties is the Gauss–Markov theorem. Many textbooks formally discuss the assumptions and proof, including Wooldridge (2010) and
Cameron and Trivedi (2005). The Gauss–Markov theorem has two main implications.
First, OLS estimates have the desirable property of being unbiased under relatively weak
conditions. Second, there is no linear estimator with better properties than OLS. These
desirable features mean, in many cases, we can use the linear regression model to estimate causal treatment effects and marginal and incremental effects of other covariates,
as we outlined in chapter 2. In this chapter, we show how these can be implemented in
Stata and discuss the interpretation of various effects.

The Gauss–Markov theorem applies to OLS models only when the assumptions are
met. If a regressor is endogenous, for example, the conditional expectation of the
error term is not zero. If it was, it would violate one of the Gauss–Markov theorem's
assumptions, and the OLS estimates would be inconsistent. With observational data,
researchers should always be aware of the possibility of endogenous regressors. We
address these issues in chapter 10.

The other main assumption is that the model specification is correct. Estimation
of a linear model without serious consideration of the model specification can lead to
substantially misleading answers. One of the most important features of any model
is the relationship of the covariates to the dependent variable. Correct specification
of the relationship is a key assumption of the theorem. In practice, while researchers
cannot claim to know the true model, they should strive to specify good models. A

Chapter 4 The linear regression model: S[

		Delta-method			
	Margin	Std. Err.	t	P>|t|	[95
_at					
1	122.1372	219.8683	0.56	0.579	-308
2	4579.065	369.6208	12.39	0.000	385
3	1264.007	149.0988	8.48	0.000	97
4	5720.936	269.5761	21.22	0.000	519
5	2963.419	125.6307	23.59	0.000	271
6	7420.347	269.2013	27.56	0.000	689
7	3428.446	106.2989	32.25	0.000	322
8	7885.374	218.5641	36.08	0.000	745
9	5804.7	344.0292	16.87	0.000	513
10	10261.63	378.7495	27.09	0.000	951
11	5592.884	277.53	20.15	0.000	504
12	10049.81	315.4738	31.86	0.000	943

```
. marginsplot
```

```
Variables that uniquely identify margins: age female anylim
```

Predicted total expenditures increase for all four types of peopl (see figure 4.1). The top two lines are for people with limitations, the lines for those people without any limitations. Women have expenditures than men at young ages, but mens' expenditures with age. Around age 70, the predictions cross; elderly men a slightly more than elderly women (controlling for limitations). 1 relationship between all the variables and shows the importance action term between age and gender.

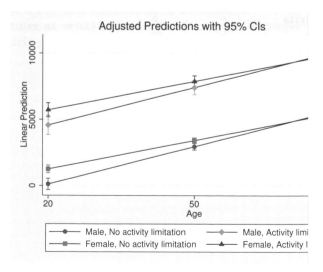

Figure 4.1. The relationship between total expenditures and women, with and without any limitations

```
. *** Treatment effect for any limitations with sample-average standard errors
. margins r.anylim, contrast(nowald)
```

```
Contrasts of predictive margins
Model VCE     : Robust
```

```
Expression    : Linear prediction, predict()
```

		Delta-method		
	Contrast	Std. Err.	[95% Conf. Interval]	
anylim (Activity limitation vs No activity limitation)	4456.928	237.8453	3990.724	4923.132

Second, in Stata, an alternative to using `margins` is estimating the ATE and the ATET using the treatment-effects command `teffects`. This command estimates ATE, ATET, and potential-outcome means based on a regression (or any common nonlinear model). Because of the importance of estimating treatment effects to our framework, we will show how to use the `teffects` command and its relationship to the results from `margins`.

Without delving too deeply into the many options available in `teffects`, we will show its basic use for a linear regression with regression adjustment. We encourage you to read Stata's *Treatment-Effects Reference Manual* entry for `teffects` to learn about other useful options, for example, inverse probability weights. The syntax for `teffects` puts the basic regression in parentheses followed by the treatment variable in parentheses. We show this in the example below.

Turning first to the ATE, we see that the treatment effect estimated by `teffects` is different from the treatment effect estimated by `margins` with `contrast()`, despite seemingly using the same model specification. The difference is several hundred dollars.

```
. *** Treatment effects ATE for any limitations
. teffects ra (exp_tot c.age##i.female) (anylim), ate

Iteration 0:   EE criterion =   4.429e-24
Iteration 1:   EE criterion =   2.198e-25

Treatment-effects estimation                    Number of obs    =      15,946
Estimator      : regression adjustment
Outcome model  : linear
Treatment model: none
```

exp_tot	Coef.	Robust Std. Err.	z	P>|z|	[95% Conf. Interval]	
ATE						
anylim						
(Activity..						
vs						
No activ..)	4238.612	250.7918	16.90	0.000	3747.069	4730.155
POmean						
anylim						
No activi..	2930.814	61.43684	47.70	0.000	2810.4	3051.228

The reason for the difference between the ATE estimated by `teffects` and the treatment effect estimated by `margins` is that the model specifications are different. The `teffects` command fits a model (not shown) in which the treatment variable is fully interacted with all covariates. It is equivalent (for the point estimates of the parameters) to running separate models for those with and without any limitations. These two methods of calculating the ATE (using `margins` or using `teffects`) will be the same if the original regression model interacts all covariates (in our example: `age`, `female`, and their interaction) with the treatment variable (`anylim`). That regression is below.

```
. *** Treatment effects ATET for any limitations
. teffects ra (exp_tot c.age##i.female) (anylim), atet

Iteration 0:    EE criterion =   4.174e-24
Iteration 1:    EE criterion =   1.149e-26

Treatment-effects estimation                    Number of obs     =      15,946
Estimator       : regression adjustment
Outcome model   : linear
Treatment model: none
```

exp_tot	Coef.	Robust Std. Err.	z	P>|z|	[95% Conf. Interval]
ATET					
anylim					
(Activity..					
vs					
No activ..)	4763.35	239.0513	19.93	0.000	4294.818 5231.882
POmean					
anylim					
No activi..	3487.452	87.96806	39.64	0.000	3315.038 3659.866

The difference between the two values of the potential-outcome means ($7,169 − $2,931 = $4,239) equals the ATE.

```
. *** Treatment effects POmeans for any limitations
. teffects ra (exp_tot c.age##i.female) (anylim), pomeans

Iteration 0:    EE criterion =   4.343e-24
Iteration 1:    EE criterion =   8.546e-26

Treatment-effects estimation                    Number of obs     =      15,946
Estimator       : regression adjustment
Outcome model   : linear
Treatment model: none
```

exp_tot	Coef.	Robust Std. Err.	z	P>|z|	[95% Conf. Interval]
POmeans					
anylim					
No activi..	2930.814	61.43684	47.70	0.000	2810.4 3051.228
Activity ..	7169.426	243.7574	29.41	0.000	6691.67 7647.181

In this section, we interpreted the results from a linear regression model. The model specification was useful for illustrative purposes, but we did not choose it through a rigorous process. Later, we will explore alternatives to linear regression for skewed positive outcomes (chapters 5 and 6), how to incorporate zeros (chapter 7), and how to control for design effects and possible endogeneity (chapters 11 and 10). However, first we show by example that misspecifying the model even in a straightforward manner can lead to inconsistency—even in the case of OLS estimates of the parameters of a linear regression. Afterward, we will show visual and statistical tests to help choose a model specification to reduce the chance of misspecification.

4.4   Consequences of misspecification

We describe two simple examples demonstrating what happens to estimates of average partial effects if a model is misspecified. In the context of those examples, we illustrate situations where the average of marginal effects is consistent, but marginal effects at specific values of covariates are inconsistent.

4.4.1   Example: A quadratic specification

Consider a specification in which there are two continuous variables—x and z—that explain y in an additive, quadratic relationship specified by

$$y_i = \alpha_0 + \alpha_1 x_i + \alpha_2 x_i^2 + \alpha_3 z_i + \alpha_4 z_i^2 + u_i$$

We artificially generate two variables—x and z—in Stata, both ranging from 2 to 8 to mimic the distribution of age (divided by 10) in our MEPS data. One variable, x, is uniformly distributed over the range, while the other variable, z, is beta$(2, 4)$ distributed, which implies that its distribution is skewed to the right. In Stata, we specify the data-generating process as

```
. *** Generate simulation data with quadratic specification
. clear
. set seed 123456
. generate x = 2 + 6*runiform()
. generate z = 2 + 6*rbeta(2,4)
. generate u = 3*rnormal()
. generate y = 1000 - 2*x + 0.3*x^2  - 2*z + 0.3*z^2 + u
```

The code for this example is available in the downloadable do-files that accompany this book.

We estimate two regressions using data drawn from this data-generating process. The first regression is misspecified, because it omits the two squared terms:

$$y_i = \beta_0 + \beta_1 x_i + \beta_2 z_i + e_i$$

The second regression is correctly specified:

$$y_i = \gamma_0 + \gamma_1 x_i + \gamma_2 x_i^2 + \gamma_3 z_i + \gamma_4 z_i^2 + u_i$$

We conducted Monte Carlo experiments using a sample size of $N = 10000$. We estimated the AMEs of x and z on y using 500 sample draws, and we summarize the results of deviations of the effects from the true values in figures 4.2 and 4.3. In each case, the solid curve represents the distribution of estimates from the correctly specified model, while the dashed curve represents the distribution of AMEs from the misspecified model. The results for the estimate of the coefficient on x suggest that the distribution of the AME of x on y is consistent even when the model is misspecified—because the distribution of x is symmetric, and the peak is near 0 deviation.

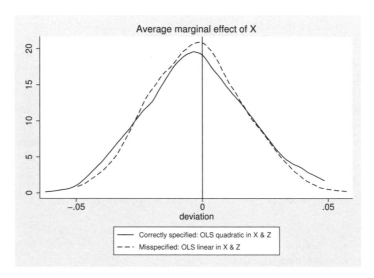

Figure 4.2. Distributions of AME of x: Quadratic specification

Figure 4.3 shows the analogous figures for the distributions of the average of marginal effects of z. The AME of z estimated from the misspecified linear-in-covariates model appears to be inconsistent. Recall that the distribution of x is symmetric, while the distribution of z is skewed. This example shows that—unless the distribution of the covariate is symmetric—even misspecification as innocuous as leaving out a quadratic term in covariates can lead to inconsistent AMEs.

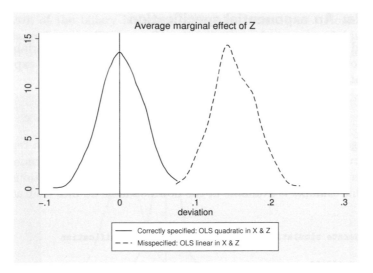

Figure 4.3. Distributions of AME of z: Quadratic specification

We now return to the distribution of the effect of x in the misspecified case. Although the distribution of the average of effects is consistent, it is important to understand its statistical properties evaluated at specific values of x. As an example, we evaluate the marginal effect of x when $x = 6$. The results, shown in figure 4.4, demonstrate that even in the case of a covariate for which the average of effects over the distribution of the covariate is consistent in the misspecified case, evidence of inconsistency appears when we evaluate the average of effects at a specified value of the covariate.

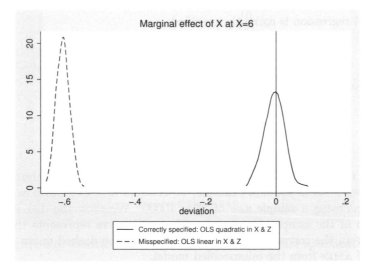

Figure 4.4. Distributions of AME of x when $x = 6$: Quadratic specification

demonstrate that the estimated effect of d on y when $d = 1$ is inconsistent when the model is misspecified as linear when the true data-generating process is exponential.

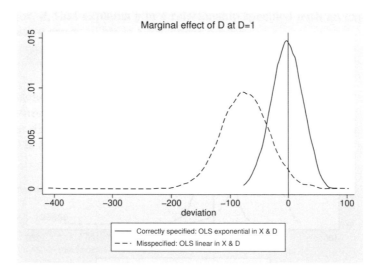

Figure 4.7. Distributions of AME of d, given $d = 1$: Exponential specification

Using those simple Monte Carlo experimental examples, we showed that the commonly held belief that effect estimates in linear models are consistent even when the model is misspecified is incorrect. In fact, the examples show that, in some cases, this belief can be grossly misleading. Thus, as we emphasize throughout the book, specification checking and testing should be an integral part of model development. We next turn to visual and statistical model checks of the linear regression specification and describe a number of ways in which specifications can be checked and improved.

4.5   Visual checks

In this section, we illustrate the use of visual residual checks for the least-squares model by examining three simple artificial-data examples with a correctly specified and a misspecified model. Then, we use the visual residual checks to explore possible misspecification in the MEPS data for two simple models.

4.5.1   Artificial-data example of visual checks

For the two artificial data examples, we draw 1,000 observations using data-generating processes, specified in Stata as

```
. *** Generate data for visual checks
. clear
. set obs 1000
number of observations (_N) was 0, now 1,000
. set seed 123456
. generate x = runiform()
. generate z = runiform()
. generate u = rnormal()
. generate y1 = 0 + 1*x + 0.0*z + u
. generate y2 = 0 + 10*x^2 + 0.0*z + u
. generate y3 = exp(-1 + 1*x + 0.0*z + u)
```

The following regression for y1 is correctly specified. Therefore, residuals from this least-squares model should have the property that there is no systematic pattern in the residuals as a function of either the predictions of y1 or the covariates.

```
. *** Fit correctly specified model: Linear in x
. regress y1 x z
```

Source	SS	df	MS		
Model	96.3935759	2	48.1967879		
Residual	1011.50979	997	1.01455345		
Total	1107.90337	999	1.10901238		

	Number of obs	=	1,000
	F(2, 997)	=	47.51
	Prob > F	=	0.0000
	R-squared	=	0.0870
	Adj R-squared	=	0.0852
	Root MSE	=	1.0073

y1	Coef.	Std. Err.	t	P>|t|	[95% Conf. Interval]	
x	1.06868	.1104417	9.68	0.000	.8519554	1.285405
z	-.1224853	.1140654	-1.07	0.283	-.346321	.1013505
_cons	.0445592	.0851434	0.52	0.601	-.1225216	.2116401

We use two **regress** postestimation commands, **rvfplot** and **rvpplot**, to detect misspecification visually. **rvfplot** plots residuals versus fit values (predicted dependent variable, or the linear index), and **rvpplot** plots residuals versus a specific predictor variable. Because this is the first time we do this, we show the Stata code for illustrative purposes.

```
. *** Create residual plots
. rvfplot, name(rvf1, replace) nodraw
. rvpplot x, name(rvpx1, replace) nodraw
. rvpplot z, name(rvpz1, replace) nodraw
. graph combine rvf1 rvpx1 rvpz1, col(3) xsize(7) ysize(2)
```

Figure 4.8 confirms our expectation that there is no pattern in the residuals. Regardless of whether we plot the residuals against predicted values, x, or z, the figures show no pattern. We conclude that the regression model is correctly specified.

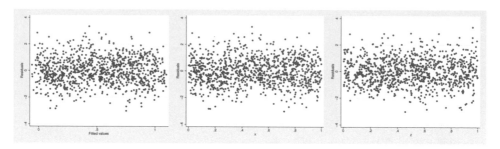

Figure 4.8. Residual plots for y1

The following regression for y2 is misspecified, because the fit model includes only a linear term. The true data-generating process is quadratic in x.

```
. *** Fit misspecified model: True model is quadratic in x
. regress y2 x z
```

Source	SS	df	MS		
Model	8303.67116	2	4151.83558	Number of obs	= 1,000
Residual	1466.09724	997	1.47050877	F(2, 997)	= 2823.40
				Prob > F	= 0.0000
				R-squared	= 0.8499
				Adj R-squared	= 0.8496
Total	9769.7684	999	9.77954795	Root MSE	= 1.2126

y2	Coef.	Std. Err.	t	P>|t|	[95% Conf. Interval]
x	9.990111	.1329626	75.13	0.000	9.729193 10.25103
z	-.0657056	.1373252	-0.48	0.632	-.3351851 .2037739
_cons	-1.612517	.1025055	-15.73	0.000	-1.813668 -1.411366

The associated residual plots in figure 4.9 show a distinct U-shaped pattern when residuals are plotted against predicted values and when they are plotted against x but show no pattern when residuals are plotted against z. Taken together, they indicate a misspecified model, likely in terms of x—but not in terms of z.

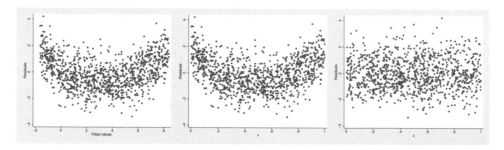

Figure 4.9. Residual plots for y2

In the third example with simulated data, we generate an outcome, y3, that is log linear in the covariates and error. Therefore, a linear regression of the logarithm of y3 would be correctly specified. However, we fit a linear regression model of y3 that is also linear in covariates.

```
. *** Fit misspecified model: True model is quadratic in x
. regress y3 x z
```

Source	SS	df	MS
Model	81.8379085	2	40.9189542
Residual	2026.66204	997	2.03276032
Total	2108.49995	999	2.11061056

Number of obs	=	1,000
F(2, 997)	=	20.13
Prob > F	=	0.0000
R-squared	=	0.0388
Adj R-squared	=	0.0369
Root MSE	=	1.4257

y3	Coef.	Std. Err.	t	P>|t|	[95% Conf. Interval]	
x	.9917468	.1563288	6.34	0.000	.6849757	1.298518
z	.0292568	.1614579	0.18	0.856	-.2875796	.3460932
_cons	.567011	.1205193	4.70	0.000	.3305104	.8035116

The associated residual plots in figure 4.10 show evidence of misspecification. Regardless of whether the residuals are plotted against predicted values, x or z, the figures show that the residuals fan out along the range of the x axis. The variation in the residuals increases with higher values of the predicted values, x and z.

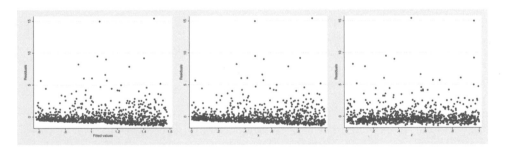

Figure 4.10. Residual plots for y3

4.5.2   MEPS example of visual checks

Are such visual representations of residuals useful in real data situations when the misspecification may not be so obvious? To demonstrate the value of such plots with real data, we construct two examples using the MEPS data for positive total expenditures and its logarithm.

```
. *** Use MEPS data; keep only positive expenditures
. use http://www.stata-press.com/data/heus/heus_mepssample, clear
(Sample of MEPS 2004 data)
. drop if exp_tot <= 0
(3,440 observations deleted)
. generate ln_exp_tot = ln(exp_tot)
```

Each of the regression specifications has two covariates, age and lninc. In the first example, we estimate a linear regression of exponentiated total expenditures on age and lninc. We construct residual plots for a 10% random sample of the MEPS observations to make the plots clearer (by reducing the density of points) and to reduce the file size of the resulting figures.

```
. *** Linear regression of expenditures on age and ln(income)
. regress exp_tot age lninc
```

Source	SS	df	MS			
				Number of obs	=	15,946
				F(2, 15943)	=	381.87
Model	8.1965e+10	2	4.0982e+10	Prob > F	=	0.0000
Residual	1.7110e+12	15,943	107320727	R-squared	=	0.0457
				Adj R-squared	=	0.0456
Total	1.7930e+12	15,945	112447746	Root MSE	=	10360

exp_tot	Coef.	Std. Err.	t	P>|t|	[95% Conf. Interval]	
age	119.0882	4.738362	25.13	0.000	109.8004	128.3759
lninc	-612.8115	89.25546	-6.87	0.000	-787.7623	-437.8608
_cons	5376.715	1020.447	5.27	0.000	3376.523	7376.907

```
. *** Create residual plots
. preserve
. set seed 123456
. keep if runiform()<0.1
(14,293 observations deleted)
. rvfplot, name(rvfm1, replace) nodraw
. rvpplot age, name(rvpxm1, replace) nodraw xlabel(20(20)80)
. rvpplot lninc, name(rvpzm1, replace) nodraw xlabel(2(2)12)
. graph combine rvfm1 rvpxm1 rvpzm1, col(3) xsize(7) ysize(2)
. restore
```

The residual plots in figure 4.11 show evidence of misspecification. The figures show that there are many small residuals below zero along the range of the x axis and a number of large, positive residuals whose frequency appears to increase from left to right along the ranges of age and lninc. Although these residual plots are intended only to be diagnostic, they do suggest that a linear-in-logs specification may be more appropriate.

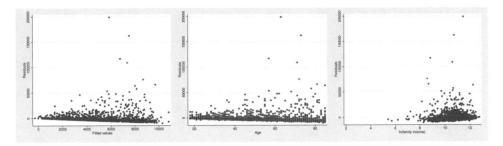

Figure 4.11. Residual plots: Regression using MEPS data, evidence of misspecification

Guided by this evidence, we generate a new dependent variable that is the natural logarithm of expenditures. Then, we estimate a linear regression of log expenditures.

```
. *** Linear regression of ln(expenditures) on age and ln(income)
. regress ln_exp_tot age lninc
```

Source	SS	df	MS		
				Number of obs =	15,946
				F(2, 15943) =	1368.77
Model	6289.40508	2	3144.70254	Prob > F =	0.0000
Residual	36628.4765	15,943	2.2974645	R-squared =	0.1465
				Adj R-squared =	0.1464
Total	42917.8816	15,945	2.69162004	Root MSE =	1.5157

ln_exp_tot	Coef.	Std. Err.	t	P>|t|	[95% Conf. Interval]	
age	.0355095	.0006933	51.22	0.000	.0341506	.0368684
lninc	-.0189931	.0130592	-1.45	0.146	-.0445906	.0066045
_cons	5.753541	.1493046	38.54	0.000	5.460888	6.046195

The residual plots in figure 4.12 show well-behaved residual scatters when plotted against predicted log expenditures and against age. However, the plot of residuals against the log of income may warrent a closer look at the specification of income. While the residuals are symmetrically distributed above and below zero along the distribution of lninc, it may be worthwhile revisiting the choice to log transform income or to distinguish between low-income observations and the rest in the regression specification.

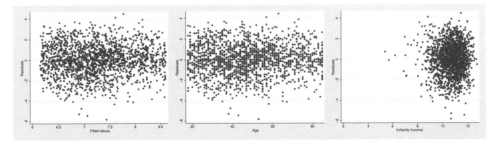

Figure 4.12. Residual plots: Regression using MEPS data, well behaved

4.6 Statistical tests

Although graphical tests are suggestive for determining model fit, they are not formal statistical tests. We present three diagnostic statistical tests that are commonly used for assessing model fit. The first two—Pregibon's (1981) link test and Ramsey's (1969) regression equation specification error test (RESET)—check for linearity of the link function. The third, a modified version of the Hosmer–Lemeshow test (1980), is a more omnibus test for model misspecification. Although each of these tests were originally developed for other applications, we focus on their interpretation as general-purpose diagnostic tests of the specification of the explanatory variables. After presenting each of these statistical tests, we then show several examples using the MEPS data.

4.6.1 Pregibon's link test

Pregibon's link test is often presented as a test of whether the independent variables are correctly specified, conditional on the specification of the dependent variable. A simple regression model shows intuition for Pregibon's link test. In a model where y is a linear function of a single regressor, x, and an intercept, we may be concerned that the model is misspecified. Alternative model specifications could include higher-order polynomials in x. The simplest alternative model is quadratic in x. Therefore, compare the model with the simple linear specification

$$y_i = \beta_0 + \beta_1 x_i + \varepsilon_i$$

with a model with a quadratic term

$$y_i = \delta_0 + \delta_1 x_i + \delta_2 x_i^2 + \eta_i$$

If the least-squares estimate of $\widehat{\delta}_2$ is significantly different from zero, we would reject the simpler model that is linear in x. The simpler specification provides an inconsistent estimate of the response to changes in x over the range of x observed in the data.

Pregibon's link test addresses the more interesting case where there are multiple covariates. For example, if there are two underlying covariates, x and z, then the quadratic expansion would include x^2, z^2, and xz. The corresponding specification test would be an F test of whether the set of estimated coefficients on the higher-order terms are statistically significantly different from zero. If there are many covariates in a model, adding all the possible higher-order terms can become unwieldy. A model with K covariates requires adding $\{K(K-1)/2\}$ additional terms.

However, we can follow similar logic by collapsing the initial model into a linear index function, $\mathbf{x}_i'\widehat{\boldsymbol{\beta}}$—which includes the constant term—and then including that index and its square in an auxiliary regression. This way, we reduce the dimensionality of the problem from K to two. Thus, replace the alternative second model with

$$y_i = \delta_0 + \delta_1 \left(\mathbf{x}_i'\widehat{\boldsymbol{\beta}}\right) + \delta_2 \left(\mathbf{x}_i'\widehat{\boldsymbol{\beta}}\right)^2 + \eta_i$$

Again, if the estimate, $\widehat{\delta}_2$, is statistically significantly different from zero, we infer that the simpler model provides an inconsistent estimate of response to changes in those covariates.

Pregibon's link test is a diagnostic test, not a constructive test. In the regression context, we cannot tell the source of the problem—missing interactions or squared terms in the covariates, a misspecification of the dependent variable, or all the above—but we can reject the simpler model specification.

4.6.2 Ramsey's RESET test

Ramsey's RESET test is a generalized version of Pregibon's link test. Although the link test works well for misspecifications that generate residual-versus-predicted plots with a U-shaped or J-shape (or the inverse), it may not work well if the plot has an S-shape or more complex shapes. These could occur if the response to the underlying covariates exhibits a threshold or diminishing returns. A quadratic formulation may not adequately reflect such a pattern.

The same logic we used to motivate Pregibon's link test can be applied to the more general RESET test. If there is a single covariate, x, we could add quadratic, cubic, and possibly quartic terms to the augmented regression. If these additional terms as a set are statistically significantly different from zero by the F test—while retaining the linear term—then we can reject the simple linear in x model in favor of a more nonlinear formulation. By extension, we can alter the link test to have a quartic alternative:

$$y_i = \delta_0 + \delta_1 \left(\mathbf{x}_i'\widehat{\boldsymbol{\beta}}\right) + \delta_2 \left(\mathbf{x}_i'\widehat{\boldsymbol{\beta}}\right)^2 + \delta_3 \left(\mathbf{x}_i'\widehat{\boldsymbol{\beta}}\right)^3 + \delta_4 \left(\mathbf{x}_i'\widehat{\boldsymbol{\beta}}\right)^4 + \eta_i$$

The RESET test is a joint F test of the hypothesis that $\widehat{\delta}_2 = \widehat{\delta}_3 = \widehat{\delta}_4 = 0$ (Ramsey 1969).

The original formulation was based on a Taylor-series expansion for a model in the spirit of what we described. The test is based on the Lagrange Multiplier principle that relaxing a constraint (under the null hypothesis that the constraint is not binding) will not improve the model fit. In some econometric textbooks, this is referred to as an omitted variable test—even though it can detect only omitted variables that are correlated with higher-order terms in the covariates. It cannot detect omitted variables that are orthogonal to included variables, such as ones that occur in randomized trials. It also cannot detect omitted variables that are linearly related to the included covariates.

4.6.3 Modified Hosmer–Lemeshow test

The link and RESET tests are parametric, because they assume that the misspecification can be captured by a polynomial of the predictions from the main equation. This alternative model specification may not be appropriate for a situation with a different pattern. For example, consider a model that was good through much of the predicted range but had a sharp downturn in the residuals when plotted against a specific covariate, x. Pregibon's link test, with its implicit assumption of a symmetric parabola, would not provide a good test for that alternative.

The modified Hosmer–Lemeshow (1980) test is a nonparametric test that looks at the quality of the predictions throughout their range. It divides the range of predictions into, for example, 10 bins. If the model specification closely approximates the data-generating process, the mean prediction for each of these bins should be near zero and not significantly different from zero.

One way to test this is to sort the data into deciles by the predicted conditional mean, $\widehat{y}_i$. Regress the residuals from the original model on indicator variables for these 10 bins, suppressing the intercept. Test whether these 10 coefficients are collectively significantly different from 0 using an F test.

The advantage of the modified Hosmer–Lemeshow test is that it is more flexible than parametric tests; the disadvantage is that it is less efficient. The modified Hosmer–Lemeshow test can be adapted to nonlinear models. Note that this test is different from the version of the Hosmer–Lemeshow test used in Stata (the `estat gof` postestimation command for logistic) to test how well data for a dichotomous outcome follows a specified dichotomous dependent variable.

4.6.4   Examples

To illustrate the link and RESET tests, we present an example using the MEPS data with a simple model specification. Consider a model that predicts total healthcare expenditures as a function only of age and gender. For this example, we drop all observations with zero expenditures:

```
. *** MEPS data, focus on positive values of total expenditures
. use http://www.stata-press.com/data/heus/heus_mepssample, clear
(Sample of MEPS 2004 data)

. drop if exp_tot <= 0
(3,440 observations deleted)
```

The model with a simple specification shows that total healthcare expenditures increase with age and are higher on average for women.

```
. *** OLS regression of exp_tot for y>0 with simple specification
. regress exp_tot age female, vce(robust)
```

Linear regression				Number of obs	=	15,946
				F(2, 15943)	=	283.26
				Prob > F	=	0.0000
				R-squared	=	0.0437
				Root MSE	=	10370

exp_tot	Coef.	Robust Std. Err.	t	P>|t|	[95% Conf. Interval]	
age	125.0642	5.26732	23.74	0.000	114.7396	135.3887
female	624.5558	173.1016	3.61	0.000	285.2571	963.8545
_cons	-1819.043	259.8691	-7.00	0.000	-2328.416	-1309.671

Because we suspect that this simple model does not accurately capture the relationship between total healthcare expenditures and demographics, we run the link and RESET tests. First, generate the predicted value of the dependent variable (linear index) and its powers up to four. In practice, it helps to normalize the linear index so the variables and parameters are neither too large nor too small. Normalization does not affect the statistical test results.

```
. *** Generate terms for link and RESET tests
. predict yhat1 if e(sample), xb

. quietly summarize yhat1

. *** Normalization
. replace  yhat1 = (yhat1 - r(min))/(r(max) - r(min))
(15,946 real changes made)

. generate yhat2 = yhat1^2

. generate yhat3 = yhat1^3

. generate yhat4 = yhat1^4
```

Pregibon's link test is a statistical test of whether the coefficient on the squared predicted value of the dependent variable is statistically different from zero in a regression of the dependent variable on the linear index and its square. The t statistic is 4.15 (corrected for heteroskedasticity) and the corresponding F statistic is 17.23 (1 and 15,943 degrees of freedom). Therefore, we conclude that the model is misspecified.

```
. *** Pregibon's link test, with vce(robust)
. regress exp_tot yhat1 yhat2, vce(robust)
```

Linear regression

				Number of obs	=	15,946
				F(2, 15943)	=	304.65
				Prob > F	=	0.0000
				R-squared	=	0.0449
				Root MSE	=	10364

exp_tot	Coef.	Robust Std. Err.	t	P>|t|	[95% Conf. Interval]	
yhat1	3645.196	1180.398	3.09	0.002	1331.484	5958.909
yhat2	5514.769	1328.681	4.15	0.000	2910.405	8119.133
_cons	1392.115	218.9845	6.36	0.000	962.8808	1821.349

```
. test yhat2

 ( 1)  yhat2 = 0

       F(  1, 15943) =    17.23
             Prob > F =     0.0000
```

Ramsey's RESET test is a statistical test of whether the coefficients on the three higher-order polynomials of the predicted value of the dependent variable are jointly statistically different from zero in a regression of the dependent variable on the fourth-order polynomial. The F statistic is 16.35 (3 and 15,941 degrees of freedom). Therefore, we conclude that the model is misspecified.

```
. *** Ramsey's RESET test, with vce(robust)
. regress exp_tot yhat1 yhat2 yhat3 yhat4, vce(robust)
```

Linear regression

				Number of obs	=	15,946
				F(4, 15941)	=	168.12
				Prob > F	=	0.0000
				R-squared	=	0.0459
				Root MSE	=	10359

exp_tot	Coef.	Robust Std. Err.	t	P>|t|	[95% Conf. Interval]	
yhat1	13890.12	4923.964	2.82	0.005	4238.598	23541.65
yhat2	-51835.46	21905.95	-2.37	0.018	-94773.59	-8897.337
yhat3	103909.3	34822.23	2.98	0.003	35653.79	172164.8
yhat4	-58068.16	18068.05	-3.21	0.001	-93483.57	-22652.75
_cons	1042.323	284.7504	3.66	0.000	484.1802	1600.466

```
. test yhat2 yhat3 yhat4

 ( 1)  yhat2 = 0
 ( 2)  yhat3 = 0
 ( 3)  yhat4 = 0

       F(  3, 15941) =    16.35
             Prob > F =     0.0000
```

There are Stata commands for both the link and RESET tests. The command for the link test is `linktest`, and the command for the RESET test is the postestimation command `estat ovtest`. However, although the `linktest` will adjust for heteroskedasticity, `estat ovtest` will not. Therefore, we do not recommend using `estat ovtest`. That is why we first showed how to calculate the test statistics without using the Stata commands. The RESET test should always be done in the manner we describe to control for possible heteroskedasticity.

For completeness, we show the results of the two Stata commands here. The `linktest` shows, numerically, the same t statistic of 4.15 as before. In contrast, the command `estat ovtest` shows a different F statistic, because it does not control for heteroskedasticity.

```
. *** Link and RESET tests with Stata commands
. quietly regress exp_tot age female, vce(robust)

. linktest, vce(robust)
```

Linear regression

				Number of obs	=	15,946
				F(2, 15943)	=	304.65
				Prob > F	=	0.0000
				R-squared	=	0.0449
				Root MSE	=	10364

exp_tot	Coef.	Robust Std. Err.	t	P>|t|	[95% Conf. Interval]	
_hat	.3460593	.1447123	2.39	0.017	.0624069	.6297118
_hatsq	.000068	.0000164	4.15	0.000	.0000359	.0001002
_cons	1229.877	273.3977	4.50	0.000	693.9867	1765.767

```
. *** vce(robust) not possible
. estat ovtest
```

Ramsey RESET test using powers of the fitted values of exp_tot
        Ho: model has no omitted variables
            F(3, 15940) =   11.83
                 Prob > F =   0.0000

The F statistic of 11.83 is smaller than the F statistic that correctly controls for heteroskedasticity, meaning the Stata command is less likely to reject the null hypothesis of correct model specification.

Next, we run the modified Hosmer–Lemeshow test on the original simple model. For the choice of 10 bins, the F test strongly rejects the null hypothesis. In Stata, we run the regression in the `nocons` mode, because OLS forces the average residual for the estimation sample to be identically zero if the original model has an intercept. We encourage researchers to use the `vce(robust)` option (and `cluster` if necessary) to correct for any heteroskedasticity.

```
. *** Modified Hosmer-Lemeshow test
. quietly regress exp_tot age female, vce(robust)

. predict yhat if e(sample), xb

. predict resid if e(sample), residual
```

```
. xtile Ipcat = yhat if e(sample), nq(10)
. sort Ipcat
. quietly regress resid ibn.Ipcat if e(sample), nocons vce(robust)
. testparm i.Ipcat
 ( 1)   1bn.Ipcat = 0
 ( 2)   2.Ipcat = 0
 ( 3)   3.Ipcat = 0
 ( 4)   4.Ipcat = 0
 ( 5)   5.Ipcat = 0
 ( 6)   6.Ipcat = 0
 ( 7)   7.Ipcat = 0
 ( 8)   8.Ipcat = 0
 ( 9)   9.Ipcat = 0
 (10)  10.Ipcat = 0

        F( 10, 15936) =     8.10
              Prob > F =   0.0000
```

Although the test clearly indicates that the specification is flawed—because the coefficients are collectively significantly different from zero—it does not provide any indications about how or why it might be flawed. We find it useful to calculate and plot the residuals at each decile of the predicted expenditure, exp_tot. These are shown in figure 4.13 along with 95% confidence intervals for those residuals shown by the dashed and dotted lines. Ideally, the graph would be flat, with all average residuals close to zero. Instead, the average residuals by decile exhibit a strong U-shape, with the lowest and middle decile residuals displaying confidence intervals outside zero.

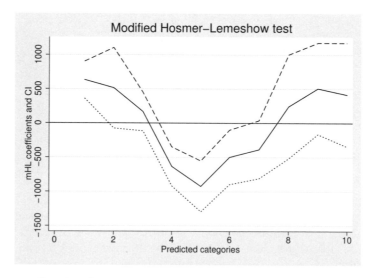

Figure 4.13. Graphical representation of the modified Hosmer–Lemeshow test

In response to the diagnostic test results, we create a richer model by adding a squared term for age and interactions between age (and age squared) with gender.

```
. *** OLS regression of exp_tot for y>0 with more general specification
. regress exp_tot c.age##c.age##i.female, vce(robust)
Linear regression                          Number of obs    =     15,946
                                           F(5, 15940)      =     122.00
                                           Prob > F         =     0.0000
                                           R-squared        =     0.0455
                                           Root MSE         =      10362
```

exp_tot	Coef.	Robust Std. Err.	t	P>|t|	[95% Conf. Interval]	
age	16.97504	36.47531	0.47	0.642	−54.5207	88.47077
c.age#c.age	1.197602	.389538	3.07	0.002	.4340638	1.961141
female Female	1797.4	1021.267	1.76	0.078	−204.3987	3799.2
female#c.age Female	−33.27691	48.40838	−0.69	0.492	−128.1628	61.60896
female# c.age#c.age Female	.1446014	.5196175	0.28	0.781	−.8739076	1.16311
_cons	260.6714	771.9373	0.34	0.736	−1252.413	1773.755

This modest elaboration of the specification shows improvement in the specification tests. The link test is no longer statistically significant ($F = 0.82$). The RESET test is still statistically significant, but the F test is now much lower at 3.93. Finally, the graphical modified Hosmer–Lemeshow test no longer exhibits a strong U-shape. Instead, only 1 of the 10 deciles (the 5th) is significantly different from 0.

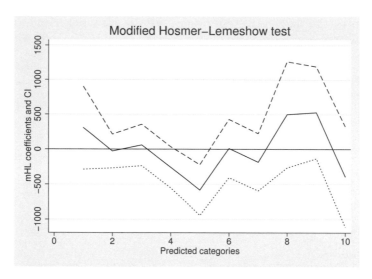

Figure 4.14. Graphical representation of the modified Hosmer–Lemeshow test after adding interaction terms

4.6.5   Model selection using Akaike information criterion and Bayesian information criterion

We now demonstrate the use of Akaike information criterion (AIC) and Bayesian information criterion (BIC) to compare model specifications (see section 2.6.1), using two of the examples from earlier in this chapter. The first examples show not only that AIC and BIC penalize models that omit important variables from the model but also that they penalize models that overfit the data by including unnecessary variables.

For the first set of examples, we use simulated data generated by a quadratic specification described in section 4.4.1. We use the same data-generating process, the same `set seed 123456`, and the same sample sizes of 1,000 and 10,000. We compare the simple specification that includes only linear terms in x and z with ones that are quadratic in x only, in z only, and one that is quadratic in both x and z. We also fit models that are cubic in x only, in z only, and one that is cubic in both x and z. There are seven models; three have omitted variables, three have extra variables, and one has the correct specification. Recall that the true specification is quadratic in both x and z, so the true specification also includes quadratic terms for both.

We label the linear specification `linear`, the quadratic ones `quadratic`, and the cubic ones `cubic`. The suffix indicates which variables are in the model.

```
. quietly regress y x z
. estimates store linear
. quietly regress y c.x##c.x c.z
```

```
. estimates store quadratic_x
. quietly regress y c.x c.z##c.z
. estimates store quadratic_z
. quietly regress y c.x##c.x c.z##c.z
. estimates store quadratic_xz
. quietly regress y c.x##c.x##c.x c.z
. estimates store cubic_x
. quietly regress y c.x c.z##c.z##c.z
. estimates store cubic_z
. quietly regress y c.x##c.x##c.x c.z##c.z##c.z
. estimates store cubic_xz
```

We report the values of the AIC and BIC from each of the models in the output shown below for the sample size of 1,000. The results show that both AIC and BIC are the smallest for the specification that is quadratic in both x and z, which is the correct specification. The linear specification has the highest values of AIC and BIC. The cubic specifications and the two misspecified quadratic specifications also have higher AIC and BIC than the correct model. Thus AIC and BIC are able to discriminate against both underspecified and overspecified models.

```
. estimates stats linear quadratic_* cubic_*
Akaike´s information criterion and Bayesian information criterion
```

Model	Obs	ll(null)	ll(model)	df	AIC	BIC
linear	1,000	-2698.945	-2572.433	3	5150.867	5165.59
quadratic_x	1,000	-2698.945	-2538.966	4	5085.932	5105.563
quadratic_z	1,000	-2698.945	-2560.713	4	5129.426	5149.057
quadratic_xz	1,000	-2698.945	-2527.777	5	5065.553	5090.092
cubic_x	1,000	-2698.945	-2538.722	5	5087.444	5111.983
cubic_z	1,000	-2698.945	-2560.576	5	5131.152	5155.691
cubic_xz	1,000	-2698.945	-2527.123	7	5068.246	5102.6

Note: N=Obs used in calculating BIC; see [R] BIC note.

When the sample size is increased to 10,000—and a new simulated sample is drawn—AIC and BIC provide even sharper contrasts between the correct specification and each of the incorrect ones.

```
. estimates stats linear quadratic_* cubic_*
Akaike´s information criterion and Bayesian information criterion
```

Model	Obs	ll(null)	ll(model)	df	AIC	BIC
linear	10,000	-26987.47	-25605.64	3	51217.29	51238.92
quadratic_x	10,000	-26987.47	-25303.79	4	50615.58	50644.43
quadratic_z	10,000	-26987.47	-25509.81	4	51027.61	51056.46
quadratic_xz	10,000	-26987.47	-25193.71	5	50397.42	50433.47
cubic_x	10,000	-26987.47	-25303.73	5	50617.45	50653.51
cubic_z	10,000	-26987.47	-25509.7	5	51029.4	51065.45
cubic_xz	10,000	-26987.47	-25193.67	7	50401.33	50451.81

```
Note: N=Obs used in calculating BIC; see [R] BIC note.
```

Next, we illustrate the use of AIC and BIC for the more realistic case of analyzing real data where the true model specification is unknown. The second set of examples compares three different model specifications for the MEPS data in section 4.4.2. Specifically, the dependent variable is total expenditures for the subsample with positive expenditures (exp_tot > 0). In the first MEPS data example, the covariates used in all three specifications are the continuous age and the binary female. They are entered additively in the first specification, an additional interaction between age and female is included in the second specification, and an additional squared age and interaction of squared age with female are included in the third specification. We label the first specification as linear, the second as interact, the richest specification with squared age, and interactions as quad_int in the output below that compares the AIC and BIC from the three models. The results show that both AIC and BIC are smallest for the specification with quadratic age and interaction terms.

```
. estimates stats linear interact quad_int
Akaike´s information criterion and Bayesian information criterion
```

Model	Obs	ll(null)	ll(model)	df	AIC	BIC
linear	15,946	-170429.4	-170072.8	3	340151.7	340174.7
interact	15,946	-170429.4	-170071.2	4	340150.3	340181
quad_int	15,946	-170429.4	-170058	6	340128	340174.1

```
Note: N=Obs used in calculating BIC; see [R] BIC note.
```

The second MEPS data example demonstrates a real-life problem that a researcher might face. So far, we have reported results of the AIC and BIC together, because the conclusions have always been the same. In this example, they are not. We add the binary `anylim` to the list of covariates, in addition to the continuous `age` and binary `female`. This variable enters additively in each of the specifications considered for the first example. AIC and BIC are calculated for each of the models and are shown in the output below. AIC is smallest for the richest specification with quadratic terms and interactions. However, BIC is smallest for the simplest additive specification.

```
. estimates stats linear interact quad_int
Akaike's information criterion and Bayesian information criterion
```

Model	Obs	ll(null)	ll(model)	df	AIC	BIC
linear	15,946	-170429.4	-169790.7	4	339589.4	339620.2
interact	15,946	-170429.4	-169787.8	5	339585.7	339624.1
quad_int	15,946	-170429.4	-169782	7	339578	339631.7

Note: N=Obs used in calculating BIC; see [R] BIC note.

The mathematical explanation is simple; the AIC penalizes models only along the number-of-parameters dimension. The BIC penalizes models using an interaction of the number of parameters and the logarithm of the sample size. So, except for very small sample sizes, the BIC will have bigger penalties for additional parameters relative to the AIC. It is a more conservative criterion. The more difficult issue is the question of what the analyst should conclude. The answer depends on whether the substantive question calls for a more parsimonious or less parsimonious regression specification.

Three additional points are worth raising. First, for nested models, we could have done other statistical tests—such as F tests. But AIC and BIC can be compared even for nonnested models, when tests such as F tests and Wald tests are not possible. Second, with standard testing procedures, the issue of multiple testing always looms large when a researcher is searching for the best model specification. AIC and BIC do not suffer from that issue. Any number of candidate models can be compared. Finally, in general, we cannot compare models with different dependent variables using AIC and BIC. In chapter 6, we demonstrate how to use AIC and BIC to compare a linear specification with one that is linear in the logged outcome.

4.7 Stata resources

The Stata command for linear regression is `regress`. The `regress` postestimation commands contain many graphic and statistical diagnostic techniques for linear regression. Diagnostic plots contain some of the commands used here as well as commands for assessing the normality and other distributional fit issues. We find `rvfplot` and `rvpplot` especially useful. The `grc1leg` command will combine graphs using one legend.

Stata has several commands to estimate and interpret the effects of covariates and treatment effects. These commands include `margins`, `margins` with the `contrast()` option, and `teffects`. Depending on whether the research question of interest is about the effect of a change in a covariate holding all other variables constant, or a population-averaged effect, these commands can get the right magnitude and standard errors. `marginsplot` is especially useful after `margins` for visualizing marginal effects and treatment effects with their standard errors.

The Stata command for the link test is `linktest`. The Stata command for the RESET test is `estat ovtest`, but this is not recommended because it does not control for heteroskedasticity. For testing hypotheses of specific variables or coefficients, use `test` and `testparm`.

To compute the AIC and BIC, use `estimates stats` after running the model. Cattaneo, Drukker, and Holland (2013) created the `bfit` command to find the model that minimizes either the AIC or the BIC from a broad set of possible models for `regress`, `logit`, and `poisson`.

For a general discussion of issues for least squares, including weighted and generalized least squares, see Christopher Baum's (2006) book, *An Introduction to Modern Econometrics Using Stata* (especially chapters 4–9). Nicholas Cox's (2004) article "Speaking Stata: Graphing model diagnostics" provides a review of the literature for diagnostic plots. He also includes a set of worked examples.

5 Generalized linear models

5.1 Introduction

As we move into the main four chapters of this book (chapters 5–8), our focus shifts toward choosing between alternative models for continuous positive outcomes, such as healthcare expenditures. One of the main themes of this book is how best to model outcomes that are not only continuous and positive but also highly skewed. Skewness creates two problems for ordinary least-squares (OLS) models: negative predictions and large sample-to-sample variation. After briefly demonstrating this for the 2004 Medical Expenditure Panel Survey (MEPS) data, we then spend the rest of this chapter exploring generalized linear models (GLM), which are an alternative to OLS that handle skewed data more easily.

As we showed in chapter 3, health expenditures are extremely skewed (see the left-panel of figure 5.1 for a density plot for total expenditures in the MEPS data for those with positive expenditures). One of the consequences of such extreme skewness is that predicted total expenditures from a linear regression model can be negative. Using the MEPS dataset, we fit a linear regression model of total healthcare expenditures (for the subsample with some use of care) on a number of covariates and then calculate predicted expenditures postestimation. We use the `centile` command to determine the fraction of negative expenditures. The results, shown below, tell us that between 6% and 7% of predictions are negative, with many substantially so.

```
. use http://www.stata-press.com/data/heus/heus_mepssample
(Sample of MEPS 2004 data)
. drop if exp_tot == . | exp_tot == 0
(3,440 observations deleted)
. quietly regress exp_tot age i.female i.race_bl i.race_oth i.eth_hisp famsize
>         i.ed_hs i.ed_hsplus i.ed_col i.ed_colplus lninc
>         i.reg_midw i.reg_south i.reg_west i.anylim mcs12 pcs12
>         i.ins_mcare i.ins_mcaid i.ins_unins ins_dent
. predict exphat
(option xb assumed; fitted values)
. centile exphat, centile(2 5 6 7)
```

Variable	Obs	Percentile	Centile	— Binom. Interp. — [95% Conf. Interval]	
exphat	15,946	2	-1316.524	-1398.257	-1221.86
		5	-389.7476	-462.5743	-315.2586
		6	-168.0266	-254.6672	-92.2117
		7	35.31625	-44.04256	105.9635

```
. predict u, residual
```

71

3. The variance, v, of the raw-scale outcome, y, is itself a function of the mean, μ, but not of the covariates, except through the mean function, $\mu\left(\mathbf{x}_i'\boldsymbol{\beta}\right)$.

4. The continuous outcome, y, is generated by a distribution from the exponential family, which includes the normal (Gaussian), Poisson, gamma, and inverse Gaussian distributions (see McCullagh and Nelder 1989).

Only certain combinations of link functions and distribution families are permitted; see McCullagh and Nelder (1989) for more details. Some popular link functions for continuous outcomes include the identity link $\{\mathbf{x}_i'\boldsymbol{\beta} = E(y_i|\mathbf{x}_i)\}$, powers $[\mathbf{x}_i'\boldsymbol{\beta} = \{E(y_i|\mathbf{x}_i)\}^{1/\delta}]$ and the natural logarithm $[\mathbf{x}_i'\boldsymbol{\beta} = \ln\{E(y_i|\mathbf{x}_i)\}]$. Common distribution families for continuous dependent variables imply variances that are integer powers of the mean function. The four most common distribution families are Gaussian, in which the variance is a constant (zero power); Poisson, in which the variance is proportional to the mean (power $= 1$); gamma, in which the variance is proportional to the square of the mean (power $= 2$); and inverse Gaussian, in which the variance is proportional to the mean cubed (power $= 3$). Table 5.1 lists commonly used links and distribution families, along with their implications for the expected value and variance of the outcome.

Table 5.1. GLMs for continuous outcomes

Common links	Link function g	Expected value
Identity	$\mathbf{x}_i'\boldsymbol{\beta} = \mu_i$	$\mu_i = \mathbf{x}_i'\boldsymbol{\beta}$
Power	$\mathbf{x}_i'\boldsymbol{\beta} = \mu_i^{1/\delta}$	$\mu_i = (\mathbf{x}_i'\boldsymbol{\beta})^{\delta}$
Log	$\mathbf{x}_i'\boldsymbol{\beta} = \ln(\mu_i)$	$\mu_i = e^{(\mathbf{x}_i'\boldsymbol{\beta})}$

Distributions	Variance	Power
Gaussian*	$v \neq f(\mu)$	0
Poisson*	$v \propto \mu$	1
Gamma*	$v \propto \mu^2$	2
Inverse Gaussian*	$v \propto \mu^3$	3

Table notes: * indicates member of the exponential family
y = raw-scale outcome
$\mathbf{x}$ = column vector of covariates, including a constant term
$\mu_i = E(y_i|\mathbf{x}_i)$ = mean of y_i conditional on $\mathbf{x}_i$
$v_i = \mathbf{V}(y_i|\mathbf{x}_i)$ = variance of y_i conditional on $\mathbf{x}_i$

The GLM approach also allows distribution families that are noninteger powers of the mean function, but such models are less common in the literature. For more details, see Blough, Madden, and Hornbrook (1999) and Basu and Rathouz (2005).

5.2.2 Parameter estimation

As alluded to above, GLM estimation requires two sets of choices. The first set of choices determines the link function and the distribution family. In section 5.8, we discuss how we chose based on rigorous statistical tests.

For the parameter estimates in the model to be consistent, it is only necessary to correctly specify the link function, g, and how the covariates enter the index function. The choice of the distribution family affects the efficiency of the estimates, but an incorrect choice does not lead to inconsistency of the parameter estimates. An inappropriate assumption about the distribution family can lead to an inconsistent estimate of the inference statistics, but this inconsistency can be remedied using robust standard errors.

The other choice is whether to estimate GLMs by quasi–maximum likelihood or iteratively reweighted least squares. In Stata, the default is quasi–maximum likelihood, which does not require specification of the full log likelihood. The choice between these two methods does not seem to matter much in practice for typical models and datasets.

After fitting a GLM, one can easily derive marginal and incremental effects of specific covariates on the expected value of y (or other treatment effects).

5.3 GLM examples

We now show how to estimate GLMs for healthcare expenditures with a few choices of link functions and distribution families, using the MEPS data introduced in chapter 3. Specifically, we estimate the effect of age (`age`) and gender (`female` is a binary indicator for being female) on total healthcare expenditures for persons with any expenditures (`exp_tot > 0`).

Our first example is a model with a log link (option `link(log)`) and a Gaussian family (option `family(gaussian)`). This is equivalent to a nonlinear regression model with an exponential mean. The results show that healthcare expenditures increase with age and are higher for women, but the coefficient on `female` is not statistically significant at the 5% level. Because the conditional mean has an exponential form, coefficients can be interpreted directly as percent changes. Expenditures increase by about 2.6% with each additional year of age after adjusting for gender. Women spend about 8% more than men $[0.080 = \exp(0.0770) - 1]$ after controlling for age.

```
. *** GLM of total expenditures, log link and Gaussian family
. glm exp_tot age female, link(log) family(gaussian) vce(robust)
Iteration 0:    log pseudolikelihood = -191206.56
Iteration 1:    log pseudolikelihood = -183815.74
Iteration 2:    log pseudolikelihood = -172074.86
Iteration 3:    log pseudolikelihood = -170297.76
Iteration 4:    log pseudolikelihood = -170068.69
Iteration 5:    log pseudolikelihood = -170067.92
Iteration 6:    log pseudolikelihood = -170067.92
```

Generalized linear models	No. of obs =	15,946
Optimization : ML	Residual df =	15,943
	Scale parameter =	1.07e+08
Deviance = 1.71351e+12	(1/df) Deviance =	1.07e+08
Pearson = 1.71351e+12	(1/df) Pearson =	1.07e+08
Variance function: V(u) = 1	[Gaussian]	
Link function : g(u) = ln(u)	[Log]	
	AIC =	21.33086
Log pseudolikelihood = -170067.9162	BIC =	1.71e+12

exp_tot	Coef.	Robust Std. Err.	z	P>|z|	[95% Conf. Interval]	
age	.0257328	.00101	25.48	0.000	.0237533	.0277123
female	.0769791	.0420732	1.83	0.067	-.0054828	.1594411
_cons	7.041415	.0652623	107.89	0.000	6.913503	7.169327

The second example also specifies a log link but assumes that the distribution family is gamma (option `family(gamma)`), implying that the variance of expenditures is proportional to the square of the mean. This is a leading choice in published models of healthcare expenditures, but we will return to the choices of link and family more comprehensively in section 5.8.

The results show that healthcare expenditures increase with age and are higher for women. Both coefficients are statistically significant, with $p < 0.001$. Expenditures increase by about 2.8% with each additional year of age, which is quite close to the effect fit by the model with the Gaussian family. However, now we find that women spend about 23% more than men ($0.23 = \exp(0.2086) - 1$), after controlling for age. This is almost three times as large as the effect estimated in the model with the Gaussian family. A small change in the model leads to a large change in interpretation.

```
. *** GLM of total expenditures, log link and gamma family
. glm exp_tot age female, link(log) family(gamma) vce(robust)

Iteration 0:   log pseudolikelihood = -150164.47
Iteration 1:   log pseudolikelihood = -148063.58
Iteration 2:   log pseudolikelihood =  -148047.8
Iteration 3:   log pseudolikelihood = -148047.79

Generalized linear models                      No. of obs        =     15,946
Optimization      : ML                         Residual df       =     15,943
                                               Scale parameter   =   7.097257
Deviance          =  33478.20882               (1/df) Deviance   =   2.099869
Pearson           =  113151.5612               (1/df) Pearson    =   7.097257

Variance function: V(u) = u^2                  [Gamma]
Link function     : g(u) = ln(u)               [Log]

                                               AIC               =   18.56902
Log pseudolikelihood =  -148047.791            BIC               =  -120801.6
```

exp_tot	Coef.	Robust Std. Err.	z	P>|z|	[95% Conf. Interval]	
age	.0279516	.0012071	23.16	0.000	.0255856	.0303176
female	.2086064	.0478756	4.36	0.000	.114772	.3024408
_cons	6.835683	.0856778	79.78	0.000	6.667757	7.003608

Our primary intent was to use these examples to demonstrate the use of the `glm` command and explain how to interpret coefficients. However, these examples also show that the estimated effects in a sample can be quite different across distribution family choices when the link function is the same, even though the choice of family has no theoretical implications for consistency of parameter estimates.

We could run many other GLM models, changing the link function or the distributional family. For example, we could fit a GLM with a square root link (option `link(power 0.5)`) and a Poisson family (option `family(poisson)`). Or we could fit a GLM with a cube root link (option `link(power 0.333)`) and an inverse Gaussian family (option `family(igaussian)`).

5.4   GLM predictions

For all GLM models with a log link, the expected value of the dependent variable, y, is the exponentiated linear index function:

$$E\left(y_i|\mathbf{x}_i\right) = \mu_i = g^{-1}\left(\mathbf{x}_i'\boldsymbol{\beta}\right) = \exp\left(\mathbf{x}_i'\boldsymbol{\beta}\right) \tag{5.1}$$

The sample average of the expected value of total expenditures is the average of μ_i over the sample. We calculate its estimate using the `margins` command. The predicted mean of total expenditures is \$4,509, less than 1% from the sample mean of \$4,480.

```
. *** Predicted mean of total expenditures from GLM with log link and gamma
> family
. quietly glm exp_tot age female, link(log) family(gamma) vce(robust)
. margins
Predictive margins                                Number of obs    =    15,946
Model VCE     : Robust
Expression    : Predicted mean exp_tot, predict()
```

		Delta-method				
	Margin	Std. Err.	z	P>|z|	[95% Conf.	Interval]
_cons	4508.963	81.41072	55.39	0.000	4349.401	4668.525

When we compare predictions from log transformation models in chapter 6 with the sample mean, we will find that those predictions are much further off. They will be anywhere from 10% to 20% too high. GLM is generally better than log models at reproducing the sample mean of the outcome.

5.5 GLM example with interaction term

Before computing marginal effects, we extend our simple specification to include the interaction of age and gender as a covariate. That is, we allow for the effect of gender to vary by age (or equivalently, the effect of age to vary by gender). The results with an interaction term are harder to interpret, but more realistic, and will help show the power of several Stata postestimation commands.

When including interaction terms, one must use special Stata notation, so that margins knows the relationship between variables when it takes derivatives. Therefore, we use c. as a prefix to indicate that age is a continuous variable, i. to indicate that female is an indicator variable, and ## between them to include not only the main effects but also their interaction.

```
. *** GLM of total expenditures, log link and gamma family
. *** Specification includes interaction between age and gender
. glm exp_tot c.age##i.female, link(log) family(gamma) vce(robust)

Iteration 0:    log pseudolikelihood = -150089.75
Iteration 1:    log pseudolikelihood = -147999.19
Iteration 2:    log pseudolikelihood = -147983.75
Iteration 3:    log pseudolikelihood = -147983.73
```

Generalized linear models		No. of obs	=	15,946
Optimization : ML		Residual df	=	15,942
		Scale parameter	=	7.567579
Deviance = 33350.09564		(1/df) Deviance	=	2.091964
Pearson = 120642.3512		(1/df) Pearson	=	7.567579
Variance function: V(u) = u^2		[Gamma]		
Link function : g(u) = ln(u)		[Log]		
		AIC	=	18.56111
Log pseudolikelihood = -147983.7344		BIC	=	-120920.1

exp_tot	Coef.	Robust Std. Err.	z	P>|z|	[95% Conf. Interval]	
age	.0345881	.0024311	14.23	0.000	.0298233	.0393529
female						
Female	.7164142	.1614244	4.44	0.000	.4000283	1.0328
female#c.age						
Female	-.0106117	.0027302	-3.89	0.000	-.0159628	-.0052607
_cons	6.513084	.1468077	44.36	0.000	6.225347	6.800822

The results are harder to interpret directly, because the interaction term allows for the effect of age to depend on gender and the effect of gender to depend on age. The coefficients on the main effects of age and gender are similar in magnitude to the simpler model. The interaction term is negative and statistically significant, implying that the increase in expenditures with age are lower for women than for men. However, to predict by how much, use `margins`.

The overall predicted total expenditure is about \$4,498 for this model, which includes age, sex, and their interaction. This is even closer to the sample mean of \$4,480 than the model without the interaction term.

```
. *** margins: Overall predicted mean
. margins
```

Predictive margins		Number of obs	=	15,946
Model VCE : Robust				
Expression : Predicted mean exp_tot, predict()				

	Margin	Delta-method Std. Err.	z	P>|z|	[95% Conf. Interval]	
_cons	4497.587	82.97491	54.20	0.000	4334.959	4660.215

If there is an interaction between two variables, x_1 and x_2, then the marginal effect with respect to x_1 will also have an extra term involving the interaction coefficient, β_{12}. This corresponds to the example with an interaction term in section 5.5.

$$\frac{\partial E\left(y_i|\mathbf{x}_i\right)}{\partial x_1} = \left(\beta_1 + \beta_{12}x_{2i}\right)\exp\left(\mathbf{x}_i'\boldsymbol{\beta}\right)$$

An incremental effect is the difference between the expected value of y_i, evaluated at two different values of a covariate of interest, holding all other covariates fixed. The formula below compares the expected value of y_i when $x_{1i} = 1$ with $x_{1i} = 0$ [and is based on (2.3)], although the specific values of x_{1i} can vary depending on the research question.

$$\frac{\Delta E\left(y_i|\mathbf{x}_i\right)}{\Delta x_1} = \exp\left(\mathbf{x}_i'\boldsymbol{\beta}|x_{1i} = 1\right) - \exp\left(\mathbf{x}_i'\boldsymbol{\beta}|x_{1i} = 0\right)$$

Incremental effects are most commonly computed for a binary covariate, like gender, whether someone has insurance or not, or if a person lives in an urban or rural area. They can also be computed for a large discrete change in a continuous variable, like age or income, which may have more policy meaning than a tiny marginal effect. We will show how to compute this kind of incremental effect in section 5.7.

When there are links other than the log link (see table 5.1), then the expectation, the marginal, and the incremental effects are based on the inverse of the link, $g^{-1}\left(\mathbf{x}_i'\boldsymbol{\beta}\right)$, and its partial derivative with respect to a specific covariate.

5.7   Example of marginal and incremental effects

Next, we compute the average marginal effects of the covariates while accounting for their interaction. Stata's `margins` command will do this correctly, as long as the relationship between the variables is indicated in the `glm` command line using the Stata symbols for variable types and its powers and interactions—namely, `c.`, `i.`, and `##`. The average marginal effect of age (averaged over men and women) is about $126. The average incremental difference between men and women (averaged over all ages) is about $508.

```
. *** Marginal effect of age and gender
. margins, dydx(age female)
Average marginal effects                    Number of obs     =      15,946
Model VCE      : Robust

Expression     : Predicted mean exp_tot, predict()
dy/dx w.r.t.   : age 1.female
```

	dy/dx	Delta-method Std. Err.	z	P>|z|	[95% Conf. Interval]	
age	126.0603	6.073027	20.76	0.000	114.1574	137.9632
female						
Female	507.6144	175.986	2.88	0.004	162.6882	852.5407

```
Note: dy/dx for factor levels is the discrete change from the base level.
```

However, the average marginal effects mask that the marginal effects vary with age. We next graph the estimated marginal effects by age and gender (see figure 5.3); this shows the slope of lines in figure 5.2. The marginal effects of age are similar for men and women until about age 40, then are higher for men at older ages.

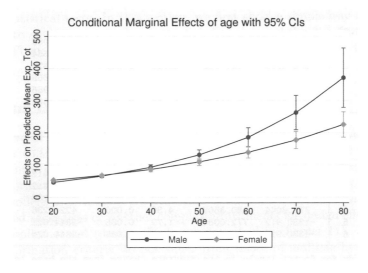

Figure 5.3. Predicted marginal effects of age by age and gender

We can use the **margins** command with the **dydx()** option to compute the marginal effects by age and gender, corresponding to figure 5.3. The results confirm what is shown in the graph. Marginal effects for men and women are similar at young ages but are much larger for men above age 60. The incremental effect of gender shows that women spend more on average at younger ages (by almost $900 at age 20), but that difference is reversed in old age, with men spending considerably more per year on average.

To illustrate the use of AIC and BIC we fit models with log and square root (power = 0.5) links using Gaussian, Poisson, and gamma distribution families. We fit six different models with these links and families using our MEPS dataset, store the results, and compare the AIC and BIC for each.

```
. *** AIC and BIC tests for link function and distribution family
. local xvar age i.female i.race_bl i.race_oth i.eth_hisp famsize
>         i.ed_hs i.ed_hsplus i.ed_col i.ed_colplus lninc
>         i.reg_midw i.reg_south i.reg_west i.anylim mcs12 pcs12
>         i.ins_mcare i.ins_mcaid i.ins_unins ins_dent

. quietly glm exp_tot `xvar´, iter(40) link(log) family(gamma)
. estimates store glm_log_gam
. quietly glm exp_tot `xvar´, iter(40) link(power .5) family(gamma)
. estimates store glm_sqrt_gam

. quietly glm exp_tot `xvar´, iter(40) link(log) family(gaussian)
. estimates store glm_log_gau
. quietly glm exp_tot `xvar´, iter(40) link(power .5) family(gaussian)
. estimates store glm_sqrt_gau

. quietly glm exp_tot `xvar´, iter(40) link(log) family(poisson) scale(x2)
. estimates store glm_log_poi
. quietly glm exp_tot `xvar´, iter(40) link(power .5) family(poisson) scale(x2)
. estimates store glm_sqrt_poi
```

Note that we also used the `scale(x2)` option for the Poisson model. This option is necessary for GLM models with a continuous dependent variable to compute correct standard errors. It is the default for Gaussian and gamma families but must be added for Poisson.

The AIC and BIC in the stored results are easily compared in a table. In our MEPS data with the full covariate specification, the model with the lowest AIC and BIC was the log link with the gamma family. Although we expected this, and this choice of link function and family often wins for expenditure data, it is always worth checking.

```
. *** Results of AIC and BIC tests
. estimates stats *

Akaike´s information criterion and Bayesian information criterion
```

Model	Obs	ll(null)	ll(model)	df	AIC	BIC
glm_log_gam	15,946	.	−144746.2	22	289536.4	289705.3
glm_sqrt_gam	15,946	.	−144839.1	22	289722.2	289891.1
glm_log_gau	15,946	.	−169126.6	22	338297.2	338466.1
glm_sqrt_gau	15,946	.	−169157.3	22	338358.5	338527.4
glm_log_poi	15,946	.	−5.00e+07	22	1.00e+08	1.00e+08
glm_sqrt_poi	15,946	.	−5.03e+07	22	1.01e+08	1.01e+08

```
Note: N=Obs used in calculating BIC; see [R] BIC note.
```

In this example, we did not fit models with the identity link. The identity link for nonnegative expenditures is both conceptually flawed and causes computational problems. The dependent variable of expenditures can never be negative, yet a model with an identity link would allow this possibility. In contrast, the log link (which exponentiates the linear index) and the square root link (which squares the linear index) never estimate the conditional mean of the dependent variable to be negative. When using the identity link with many datasets (including the MEPS example), a rich set of covariates will predict the conditional mean to be negative for some observations. For these observations, and hence for the sample as a whole, the log-likelihood function is undefined. In such cases, the maximum likelihood estimation will have trouble finding a solution. For other types of dependent variables, the identity link function may well be appropriate. As a precaution, in our empirical example, we use the `iter(40)` option to limit the number of iterations to be 40, so that it will not iterate forever. Typically, GLMs converge in less than 10 iterations. Consequently, if the model gets to 40 iterations, check to see if there is a problem with the model as specified.

5.8.2 Test for the link function

Instead of choosing both the link function and the distribution family simultaneously, choose them sequentially using a series of statistical tests. Use a Box–Cox approach (see section 6.5) to find an appropriate functional form and use that form as the link function. In brief, the Box–Cox approach tests which scalar power, δ, of the dependent variable, y^{δ}, results in the most symmetric distribution. A power of $\delta = 1$ corresponds to a linear model, $\delta = 0.5$ corresponds to the square root transformation, and $\delta = 0$ corresponds to the natural log transformation model. This approach is discussed at length in section 6.5, with examples that show that the log link is preferred to the square root for the MEPS dataset and the basic model.

Note that the `boxcox` command does not admit the factor-variable syntax of modern Stata. Therefore, we use the `xi:` prefix to preprocess the data to generate appropriate indicators. The estimated coefficient (`/theta` in the output) is only slightly greater than zero. We take this to mean that the log link function is preferable to the square root or other common link functions.

```
. *** Link test
. xi: boxcox exp_tot age i.female i.race_bl i.race_oth i.eth_hisp famsize
>           i.ed_hs i.ed_hsplus i.ed_col i.ed_colplus lninc
>           i.reg_midw i.reg_south i.reg_west i.anylim mcs12 pcs12
>           i.ins_mcare i.ins_mcaid i.ins_unins ins_dent
i.female          _Ifemale_0-1          (naturally coded; _Ifemale_0 omitted)
i.race_bl         _Irace_bl_0-1         (naturally coded; _Irace_bl_0 omitted)
i.race_oth        _Irace_oth_0-1        (naturally coded; _Irace_oth_0 omitted)
i.eth_hisp        _Ieth_hisp_0-1        (naturally coded; _Ieth_hisp_0 omitted)
i.ed_hs           _Ied_hs_0-1           (naturally coded; _Ied_hs_0 omitted)
i.ed_hsplus       _Ied_hsplus_0-1       (naturally coded; _Ied_hsplus_0 omitted)
i.ed_col          _Ied_col_0-1          (naturally coded; _Ied_col_0 omitted)
i.ed_colplus      _Ied_colplu_0-1       (naturally coded; _Ied_colplu_0 omitted)
i.reg_midw        _Ireg_midw_0-1        (naturally coded; _Ireg_midw_0 omitted)
i.reg_south       _Ireg_south_0-1       (naturally coded; _Ireg_south_0 omitted)
i.reg_west        _Ireg_west_0-1        (naturally coded; _Ireg_west_0 omitted)
i.anylim          _Ianylim_0-1          (naturally coded; _Ianylim_0 omitted)
i.ins_mcare       _Iins_mcare_0-1       (naturally coded; _Iins_mcare_0 omitted)
i.ins_mcaid       _Iins_mcaid_0-1       (naturally coded; _Iins_mcaid_0 omitted)
i.ins_unins       _Iins_unins_0-1       (naturally coded; _Iins_unins_0 omitted)
Fitting comparison model

Iteration 0:    log likelihood = -170429.36
Iteration 1:    log likelihood = -146003.52
Iteration 2:    log likelihood = -145803.08
Iteration 3:    log likelihood = -145803.04
Iteration 4:    log likelihood = -145803.04

Fitting full model

Iteration 0:    log likelihood = -169285.62
Iteration 1:    log likelihood = -144474.58
Iteration 2:    log likelihood = -142698.56
Iteration 3:    log likelihood = -142698.06
Iteration 4:    log likelihood = -142698.06

                                        Number of obs   =      15,946
                                        LR chi2(21)     =     6209.95
Log likelihood = -142698.06             Prob > chi2     =       0.000
```

exp_tot	Coef.	Std. Err.	z	P>|z|	[95% Conf. Interval]	
/theta	.0644479	.0038325	16.82	0.000	.0569364	.0719594

Estimates of scale-variant parameters

	Coef.
Notrans	
age	.0250348
_Ifemale_1	.4647753
_Irace_bl_1	-.4131732
_Irace_oth_1	-.701529
_Ieth_hisp_1	-.5203265
famsize	-.0829778
_Ied_hs_1	.1654191
_Ied_hsplu~1	.3364049
_Ied_col_1	.4534274
_Ied_colpl~1	.5493796
lninc	.1204196
_Ireg_midw_1	-.0284099
_Ireg_sout~1	-.199333
_Ireg_west_1	-.2572476
_Ianylim_1	.5349887
mcs12	-.0248666
pcs12	-.0650513
_Iins_mcar~1	.366532
_Iins_mcai~1	.2687533
_Iins_unin~1	-1.116425
ins_dent	.2753434
_cons	10.96559
/sigma	2.141619

Test HO:	Restricted log likelihood	LR statistic chi2	P-value Prob > chi2
theta = -1	-181225.05	77053.97	0.000
theta = 0	-142840.12	284.11	0.000
theta = 1	-169285.62	53175.13	0.000

5.8.3   Modified Park test for the distribution family

There is a regression-based statistical test based on Park (1966)—called the modified Park test—that provides a simple way to test for the relationship between the predicted mean and variance in a GLM. The selection of the distribution family is important, because it affects the precision of the estimated response, both in terms of estimated coefficients and marginal effects (Manning and Mullahy 2001). In the absence of any guidance from theory, the analyst must determine empirically how the raw-scale variance depends on the mean function.

To implement the modified Park test, we first run a GLM—which means choosing an initial link function and distribution family prior to running the empirical test. Our working assumption—based on results in section 5.8 and in the literature—is that we should have a log link and gamma family. Note that this test requires link to be correctly specified. Postestimation, we generate the log of the squared residuals and the linear index.

```
. *** Run GLM to generate residuals for Park test
. quietly glm exp_tot age i.female i.race_bl i.race_oth i.eth_hisp famsize
>         i.ed_hs i.ed_hsplus i.ed_col i.ed_colplus lninc
>         i.reg_midw i.reg_south i.reg_west i.anylim mcs12 pcs12
>         i.ins_mcare i.ins_mcaid i.ins_unins ins_dent, link(log) family(gamma)
. *** Generate ln(raw residuals squared) and xbetahat for Park test
. predict double rawresid, response
. generate lnrawresid2 = ln(rawresid^2)
. predict double xbetahat, xb
```

The Park test is based on the estimated relationship between the variance of the error term and mean. The dependent variable is the natural logarithm of the raw-scale variance. The sole covariate is the natural log of the conditional expected value of the dependent variable, $\mathbf{x}_i'\widehat{\boldsymbol{\beta}} = \ln\{E(y_i|\mathbf{x}_i)\}$. The coefficient on the predicted value indicates which distribution family is preferred. Typically, analysts have made choices that reflect integer powers:

- If it is close to zero, use the Gaussian family, because the variance is unrelated to the mean, as in Mullahy (1998).

- If it is close to one, use the Poisson, because of its property that the variance is proportional to the mean.

- If it is close to two, use the gamma, as in Blough, Madden, and Hornbrook (1999).

- If it is close to three, use the inverse Gaussian.

Although the test rejects each of the integer-valued powers considered, the estimated coefficient on xbetahat (1.82) is closest to the gamma family's integer value of 2. Therefore, for these data and this model specification, we choose the gamma family (given the choice of the log link function).

```
. *** Modified Park test
. regress lnrawresid2 xbetahat, robust
```

Linear regression				Number of obs	=	15,946
				F(1, 15944)	=	9371.32
				Prob > F	=	0.0000
				R-squared	=	0.3536
				Root MSE	=	1.9819

lnrawresid2	Coef.	Robust Std. Err.	t	P>|t|	[95% Conf. Interval]	
xbetahat	1.820219	.0188028	96.81	0.000	1.783363	1.857075
_cons	.4211818	.1519263	2.77	0.006	.1233892	.7189744

5.8.4   Extended GLM

What if the appropriate link function is not one of the widely used choices [identity, square root, or $\ln(y)$]? What if the distribution family is not an integer power of the mean function? Basu and Rathouz (2005) address these questions with an approach known as the extended estimating equations model. They simultaneously estimate the mean and distribution family, rather than separately, and allow for general noninteger choices of the power values.

5.9   Conclusions

In summary, GLM is appealing in health economics, because it deals with skewness and heteroskedasticity while avoiding retransformation issues of OLS models with a logged dependent variable (see chapter 6). As we will demonstrate, it is also much harder to calculate marginal effects for log transformation models than for GLM models.

5.10   Stata resources

To estimate GLM models in Stata, use the `glm` command, which works with `margins`, `svy`, and `bootstrap`. Basu and Rathouz (2005) have Stata code for their extended estimating equations model.

To estimate predicted values and calculate marginal and incremental effects of covariates conditional on the other covariates, use the postestimation commands `margins` or `contrast` after fitting a GLM model. The `marginsplot` command generates graphs immediately after using `margins`.

To compare the AIC and BIC test statistics for GLMs with different choices of link function and distribution family, use `estimates stats *`. Alternatively, conduct a link test with `boxcox` and a modified Park test with code found in this chapter.

6 Log and Box–Cox models

6.1 Introduction

Despite the ease of fitting and interpreting generalized linear models (GLM) (see chapter 5) and the ability of GLMs to deal with heteroskedasticity while avoiding retransformation problems, a sizable fraction of the health economics literature still fits regression models with a logged dependent variable. In this chapter, we cover log models in detail to show their weaknesses and to explain how a careful analysis would properly interpret results.

Interpreting the effects of covariates on the raw scale is much more difficult than fitting log models. We are typically not interested in log dollars per se (Manning 1998). Instead, the interest ultimately is about $E(y_i|\mathbf{x}_i)$ or how $E(y_i|\mathbf{x}_i)$ changes with changes in covariates. Transforming the dependent variable for estimation complicates interpretation on the raw (unlogged) scale. Retransforming the results back to the raw scale requires dealing with the error term—which may be nonnormal, heteroskedastic, or both. Predicted values of the dependent variable and marginal effects therefore depend not only on the coefficients but also on the distribution of the error term.

In this chapter, we focus on fitting and interpreting models of transformed positive dependent variables. We start with the popular log model and later discuss the more general Box–Cox model. After introducing the model with a logged dependent variable, we explain why retransformation is difficult and dependent on the error term. We then explain how to compute $E(y_i|\mathbf{x}_i)$, marginal effects, and incremental effects under four different assumptions about the error term (homoskedastic or heteroskedastic and normal or nonnormal). Because of the importance of comparing different kinds of models, we show how to compare ordinary least-squares (OLS) regression models with either y or $\ln(y)$ as the dependent variable and then discuss in detail the differences between log models and GLM models. In the remainder of this chapter, we describe a more general transformation-based model, the Box–Cox model (1964).

6.2 Log models

6.2.1 Log model estimation and interpretation

We start with models that take the natural logarithm of a continuous dependent variable y, with no zeros or negative values (for models that include zero values, see chapter 7). For notational simplicity throughout this chapter, we assume $y > 0$ unless otherwise specified.

When the dependent variable, y, for observation i is transformed by taking the natural logarithm, the model is

$$\ln (y_i) = \mathbf{x}'_i \boldsymbol{\beta} + \varepsilon_i \qquad (6.1)$$

where $\mathbf{x}$ is a vector of covariates including the constant term, $\boldsymbol{\beta}$ is the vector of parameters to be estimated, and ε is a random error term.

The expected value of the natural logarithm of y_i (conditional on $y_i > 0$ and on $\mathbf{x}_i$) is the linear index

$$
\begin{aligned}
E \left\{ \ln (y_i) \, | \mathbf{x}_i \right\} &= E \left(\mathbf{x}'_i \boldsymbol{\beta} + \varepsilon_i | \mathbf{x}_i \right) \\
&= \mathbf{x}'_i \boldsymbol{\beta}
\end{aligned}
$$

when the error term satisfies the orthogonality constraint that $E(\mathbf{x}'\varepsilon) = 0$.

In this example, we use the 2004 Medical Expenditure Panel Survey (MEPS) data introduced in chapter 3 to estimate the effect of age (`age`) and gender (`female` is a binary indicator for being female) on total healthcare expenditures for persons with any expenditures (`exptot > 0`). We fit the log transformation model once, then use those results throughout this chapter to make predictions about y (on the raw scale), given different assumptions about the heteroskedasticity and normality of the errors.

The OLS regression results show that the log of healthcare expenditures increases with age and is higher for women. For a semilog model, it is easy to interpret the coefficient on a continuous variable—like age—as a percent change in the dependent variable. Expenditures increase by about 3.6% with each additional year of age among those who spent anything. In this case, a coefficient of 0.0358 corresponds to about a 3.6% increase, because the parameter is close to 0. A more precise value is found by exponentiating the coefficient; this more precise mathematical formula matters more for coefficients further from zero. The coefficient is statistically significantly different from 0, with $p < 0.001$.

```
. use http://www.stata-press.com/data/heus/heus_mepssample
(Sample of MEPS 2004 data)

. *** OLS regression of ln(y) for y>0
. drop if exp_tot <= 0     // focus on positive values of y
(3,440 observations deleted)

. generate ln_exp_tot = ln(exp_tot)
```

```
. regress ln_exp_tot age female, robust
Linear regression                                    Number of obs   =     15,946
                                                     F(2, 15943)     =    1639.48
                                                     Prob > F        =     0.0000
                                                     R-squared       =     0.1575
                                                     Root MSE        =      1.506
```

ln_exp_tot	Coef.	Robust Std. Err.	t	P>|t|	[95% Conf. Interval]	
age	.0358123	.0006509	55.02	0.000	.0345365	.0370881
female	.3511679	.0242947	14.45	0.000	.3035476	.3987883
_cons	5.329011	.0369364	144.28	0.000	5.256611	5.40141

Dummy variables also have a percentage effect on the dependent variable in a log model (Halvorsen and Palmquist 1980). The magnitude of the percentage change in y for a unit change in the dummy variable, x_d, is the exponentiated coefficient, $\widehat{\beta}_d$, less 1, multiplied by 100.

$$\text{Percent change in } y = 100 \times \left(e^{\widehat{\beta}_d} - 1 \right) \tag{6.2}$$

The coefficient on female in the simple health expenditure model is 0.35, meaning that women spend about 42% more than men: $42 = 100 \times \{\exp(0.35) - 1\}$, averaged over all ages.

However, (6.2) has finite sample bias, because $\widehat{\beta}_d$ is estimated with error—and because $E(e^{\widehat{\beta}_d}) \neq e^{E(\widehat{\beta}_d)}$ [see (6.3)]. Kennedy (1981) proposed subtracting a term in the exponent to correct the bias in (6.2). The Kennedy transformation is the following formula,

$$\text{Percent change in } y = 100 \times \left(e^{\{\widehat{\beta}_d - 0.5\widehat{V}(\widehat{\beta}_d)\}} - 1 \right)$$

where $\widehat{V}\left(\widehat{\beta}_d\right)$ is the OLS estimate of the variance of the coefficient on the dummy variable of interest. This formula applies only to positive coefficients. For a negative coefficient, redefine the variable by taking one minus the variable.

The variance, $\widehat{V}$, of the percentage change in y is also easy to calculate (van Garderen and Shah 2002):

$$\widehat{V} \left(\text{Percent change in } y\right) = 100^2 \times e^{2\widehat{\beta}_x} \left(e^{-\widehat{V}(\widehat{\beta}_d)} - e^{-2\widehat{V}(\widehat{\beta}_d)} \right)$$

However, in the MEPS example the Kennedy transformation is not substantively different from the standard interpretation. See the output below, which calculates both—along with the standard error—based on the formula by van Garderen and Shah (2002).

```
. *** Kennedy transformation and standard error
. quietly regress ln_exp_tot age female
. matrix V = e(V)
. scalar Vc = V[rownumb(V,"female"),rownumb(V,"female")]
. display "%Change = " 100*(exp(_b[female]) - 1)
%Change = 42.072589
. display "Kennedy = " 100*(exp(_b[female] - .5*(Vc)) - 1)
Kennedy = 42.030807
. display "s.e.    = " 100*sqrt(exp(2*_b[female])*(exp(-Vc)-exp(-2*Vc)))
s.e.    = 3.4443421
```

In our experience, the Kennedy transformation, while popular, rarely makes a substantive difference for statistically significant parameters. In the MEPS example, because the variance of $\widehat{\beta}_d$ is small, the Kennedy transformation does not make a practical difference.

6.3   Retransformation from ln(y) to raw scale

When the dependent variable y_i is transformed by taking the natural logarithm [as in (6.1)], then the expected value of y_i, conditional on $\mathbf{x}_i$, is not simply the exponentiated linear index:

$$E(y_i|\mathbf{x}_i) \neq e^{\mathbf{x}_i'\boldsymbol{\beta}} \tag{6.3}$$

Instead, it also depends on the expected value of the exponentiated error term.

$$
\begin{aligned}
E(y_i|\mathbf{x}_i) &= E\left(e^{\mathbf{x}_i'\boldsymbol{\beta}+\varepsilon_i}|\mathbf{x}_i\right) \\
&= E\left(e^{\mathbf{x}_i'\boldsymbol{\beta}}|\mathbf{x}_i\right) \times E\left(e^{\varepsilon_i}|\mathbf{x}_i\right) \\
&= e^{\mathbf{x}_i'\boldsymbol{\beta}} \times E\left(e^{\varepsilon_i}|\mathbf{x}_i\right)
\end{aligned}
\tag{6.4}
$$

The expected value of the exponentiated error term [the second term in (6.4)] is greater than one by Jensen's inequality, implying that the exponentiated linear index (6.3) is an underestimate of the expected value of y_i.

The following subsection describes two ways to estimate the multiplicative retransformation factor, $E(e^{\varepsilon_i}|\mathbf{x}_i)$, depending on whether the log scale error, ε_i, has a normal or nonnormal distribution. We then show how to calculate predicted values and their standard errors and compare the predicted values of y using each method.

6.3.1   Error retransformation and model predictions

We wrote a small program to estimate the retransformation factors using normal theory and Duan's smearing retransformation. This allows us to use the bootstrap to obtain appropriate standard errors for predicted means. As the code below shows, we first fit a linear model for $\ln(y)$ and predict the linear index (xbhat), the residuals (ehat), and the exponentiated index (expxbhat).

The first—and simplest—case assumes that the error term ε has a normal distribution. In this case, the error retransformation factor is $E(e^{\varepsilon_i}|\mathbf{x}_i) = \exp\left(0.5\sigma^2\right)$, where σ^2 is the variance of the error term on the log scale. The expected value of y_i, conditional on $\mathbf{x}_i$, is the exponentiated linear index multiplied by the error retransformation factor.

$$E(y_i|\mathbf{x}_i) = e^{\mathbf{x}_i'\boldsymbol{\beta}} \times e^{0.5\sigma^2} \tag{6.5}$$

In the program below, the normal factor is denoted normalfactor.

In the second case, we relax the normality assumption. Duan (1983) developed a consistent way to estimate $E(e^{\varepsilon_i}|\mathbf{x}_i)$ when the errors are not normal but with the covariates assumed fixed. Duan's smearing factor—denoted D_{smear}—is the scalar average of the exponentiated estimated error terms $\widehat{\varepsilon}_i = \ln(y_i) - \mathbf{x}_i'\widehat{\boldsymbol{\beta}}$, where the log-scale residual provides a consistent estimator for the error.

$$D_{\mathrm{smear}} = \frac{1}{N} \sum_{i=1}^{N} e^{\widehat{\varepsilon}_i}$$

The expected value of y_i, conditional on $\mathbf{x}_i$, is the exponentiated linear index multiplied by Duan's smearing factor:

$$E\left(y_i|\mathbf{x}_i\right) = e^{\mathbf{x}_i'\boldsymbol{\beta}} \times D_{\mathrm{smear}}$$

In the program below, Duan's smearing factor is denoted duanfactor. In each case, the predicted conditional mean is calculated by multiplying expxbhat with the appropriate multiplicative factor.

```
    . program lnmodelpredictions, eclass
    1.       capture drop xbhat ehat expxbhat normalfactor duanfactor yhat_*
    2.       *** Store ln(y) results for later use
    .        quietly regress ln_exp_tot age female
    3.       estimates store lnymodel
    4.       predict xbhat, xb
    5.       predict ehat, residual
    6.       generate expxbhat = exp(xbhat)
    7.       *** Normal factor
    .        egen normalfactor = mean(ehat^2)
    8.       replace normalfactor = exp(0.5*normalfactor)
    9.       generate yhat_normalfactor = expxbhat * normalfactor
   10.       *** Duan factor
    .        egen duanfactor = mean(exp(ehat))
   11.       generate yhat_duanfactor = expxbhat * duanfactor
   12.       mean normalfactor duanfactor exp_tot yhat_*
   13. end
```

We use the bootstrap command to calculate standard errors for the smearing factors and for the predicted means of exp_tot. We have used 200 bootstrap replications without experimentation in this example. We urge readers to ensure that the estimates of interest are stable given the choice of number of replications. We could also have used GMM to obtain correct standard errors analytically. We provide an example of how to use GMM in section 10.4.

The results show that the estimated value of the normal retransformation factor is 3.11. Clearly, 3.11 is far greater than 1.0. The estimate of Duan's smearing factor is 2.89 in the MEPS sample. Therefore, all predictions based on the normal theory retransformation factor will be 7.6% higher ($1.076 = 3.11/2.89$) than the corresponding predictions based on Duan's smearing factor. Note that the confidence intervals for these estimates do not overlap.

We now compare the sample averages of the predicted values of exp_tot with each other and with the sample average of exp_tot. The mean of total healthcare expenditures for this sample with positive expenditures is $4,480. Ideally, predictions of total healthcare expenditures on the raw scale would be close to the actual sample mean. However, the method that assumes normality does not come close—the prediction of the average is nearly 20% higher than the sample mean, and their confidence intervals do not overlap. Relaxing the normality assumption yields a prediction ($4,986) that is still too high by about 10% relative to the sample mean. Once again, the confidence intervals do not overlap. Therefore, none of these methods came particularly close to estimating the overall mean (unlike the GLM model with a log link, as in chapter 5).

```
. bootstrap _b, reps(200) seed(123456): lnmodelpredictions
(running lnmodelpredictions on estimation sample)

Bootstrap replications (200)
────┼─── 1 ───┼─── 2 ───┼─── 3 ───┼─── 4 ───┼─── 5
..................................................      50
..................................................     100
..................................................     150
..............................................        200

Mean estimation                    Number of obs   =     15,946
                                   Replications    =        200
```

	Observed Mean	Bootstrap Std. Err.	Normal-based [95% Conf. Interval]	
normalfactor	3.107321	.0406722	3.027605	3.187037
duanfactor	2.887938	.0618077	2.766797	3.009078
exp_tot	4480.262	83.63575	4316.339	4644.185
yhat_normalfactor	5364.343	90.51575	5186.935	5541.75
yhat_duanfactor	4985.609	124.7519	4741.099	5230.118

We do not expect these alternatives to have a mean exactly equal to the sample mean. However, when the predictions are far from the sample mean, there could be problems with the retransformation, the model specification, the estimate of the retransformation factor, the modeling of heteroskedasticity, or all the above. In particular, adding more covariates would improve the estimate greatly. We have also not considered heteroskedasticity. Introducing heteroskedasticity into the retransformation factor matters greatly. Simple groupwise heteroskedasticity can be easily introduced into the normal and Duan retransformation factors. Unfortunately, it is rare for a researcher to encounter situations where heteroskedasticity occurs only by group or for the researcher to be able to identify such groups in the data. Therefore, in such cases, GLMs have a natural advantage.

6.3.2 Marginal and incremental effects

Retransformation also affects estimates of the marginal effects of continuous covariates x_k. The general formula for the marginal effect applies the chain rule to the derivative of $E(y_i|\mathbf{x}_i) = e^{\mathbf{x}_i'\boldsymbol{\beta}} \times E(e^{\varepsilon_i}|\mathbf{x}_i)$:

$$\frac{\partial E\left(y_i|\mathbf{x}_i\right)}{\partial x_k} = \frac{\partial e^{\mathbf{x}_i'\boldsymbol{\beta}}}{\partial x_k} \times E\left(e^{\varepsilon_i}|\mathbf{x}_i\right) + e^{\mathbf{x}_i'\boldsymbol{\beta}} \times \frac{\partial E\left(e^{\varepsilon_i}|\mathbf{x}_i\right)}{\partial x_k} \tag{6.6}$$

The second term in (6.6) depends on whether the error term is homoskedastic or heteroskedastic. If the error term is homoskedastic, the second term is identically zero. However, ignoring heteroskedasticity—if it exists—will lead to inconsistent estimates not only of the conditional mean but also of the marginal effects (Manning 1998). This point is worth emphasizing: unlike OLS models without retransformation, heteroskedasticity must be accounted for when fitting marginal effects in log transformation models.

As with marginal effects, the calculation of incremental effects depends on correctly modeling possible heteroskedasticity, because the heteroskedasticity correction will appear in the estimate of the conditional mean of y.

6.4 Comparison of log models to GLM

There is often confusion between GLM with a log link function (see chapter 5) and OLS regression with a log-transformed dependent variable (as described in this chapter).

- GLM with a log link function models the logarithm of the expected value of y, conditional on $\mathbf{x}$—that is, $\ln\{E(y_i|\mathbf{x}_i)\}$.

- OLS regression with a log-transformed dependent variable models the expected value of the logarithm of y conditional on $\mathbf{x}$—that is, $E\{\ln(y_i)|\mathbf{x}_i\}$.

The similarity is deceptive, but the order of operations matters greatly. We compare the equations to show why these models differ.

A GLM with a log link models the log of the expected value of y, conditional on $\mathbf{x}$ as a linear index of covariates $\mathbf{x}$ and parameters $\boldsymbol{\beta}^{\text{GLM}}$:

$$\ln \left\{ E \left(y_i | \mathbf{x}_i \right) \right\} = \mathbf{x}_i' \boldsymbol{\beta}^{\text{GLM}} \tag{6.7}$$

Exponentiating (6.7) yields an expression for the expected value of y, conditional on $\mathbf{x}$:

$$E \left(y_i | \mathbf{x}_i \right) = \exp \left(\mathbf{x}_i' \boldsymbol{\beta}^{\text{GLM}} \right) \tag{6.8}$$

In contrast, an OLS regression with a log-transformed dependent variable models the log of y as a linear index of covariates $\mathbf{x}$ and parameters $\boldsymbol{\beta}^{\ln y}$, plus an error term. Notice the inclusion of an error term ε:

$$\ln \left(y_i \right) = \mathbf{x}_i' \boldsymbol{\beta}^{\ln y} + \varepsilon_i \tag{6.9}$$

Taking the expected value of both sides of (6.9) eliminates the mean-zero error term, but the resulting equation is in terms of the expected value of the logarithm of y, not the expected value of y:

$$E \left\{ \ln \left(y_i \right) | \mathbf{x}_i \right\} = \mathbf{x}_i' \boldsymbol{\beta}^{\ln y}$$

Equation (6.9) differs from (6.7) on the left-hand side, because the order of operations is different—and it differs on the right-hand side, because the parameter values are different.

To get an expression in terms of the expected value of y, return to (6.9) and first exponentiate both sides, then take the expectation. The expected value of y, conditional on $\mathbf{x}$, is therefore a complicated function of the exponentiated error term:

$$
\begin{aligned}
E \left(y_i | \mathbf{x}_i \right) &= E \left(e^{\mathbf{x}_i' \boldsymbol{\beta}^{\ln y} + \varepsilon_i} | \mathbf{x}_i \right) \\
&= \left(e^{\mathbf{x}_i' \boldsymbol{\beta}^{\ln y}} \right) \times E \left(e^{\varepsilon_i} | \mathbf{x}_i \right)
\end{aligned}
\tag{6.10}
$$

The expected value of the dependent variable, y, in the log transformation model depends on two terms [see (6.10)]. The first term looks like the expected value of y in the GLM with a log link [right-hand side of (6.8)]. However, the second term is different. It depends on the error term. If the error term is heteroskedastic in $\mathbf{x}$, the second term will also include terms in $\mathbf{x}$.

In general, the parameters from these two models will not be equal (that is, $\boldsymbol{\beta}^{\ln y} \neq \boldsymbol{\beta}^{\text{GLM}}$). In particular, the constant terms will be quite different—with $\beta_0^{\ln y} < \beta_0^{\text{GLM}}$— because in the log transformation model, $E \{ \exp(\varepsilon_i) | \mathbf{x}_i \} > 1$.

In summary, while OLS regressions with a log-transformed dependent variable appear similar to GLM models with a log link, the GLM models are easier to interpret on the raw scale and naturally adjust for heteroskedasticity. Properly interpreting results from a log-transformation model requires substantially more effort.

6.5 Box–Cox models

The log transformation is a specific case of the popular Box–Cox transformation. The Box–Cox (1964) transformation is a nonlinear transformation of a variable using a power function. Specifically, the Box–Cox transformation for $y > 0$ is a specific variant of

$$y^{(\lambda)} = \frac{y^\lambda - 1}{\lambda}$$

where $y^{(\lambda)} = \ln(y)$ in the limit as $\lambda \longrightarrow 0$. One reason for the popularity of the Box–Cox model is that it incorporates many commonly used models—including linear square root, and natural logarithm. Below are common values of λ and the corresponding power functions.

Table 6.1. Box–Cox formulas for common values of λ

Model name	λ	$y^{(\lambda)}$
Linear	1	$y - 1$
Square root	0.5	$2\left(\sqrt{y} - 1\right)$
Natural logarithm	0	$\ln(y)$
Inverse	-1	$1 - \frac{1}{y}$

By choosing the correct value of λ, we find that the resulting transformed continuous dependent variable will be closer to being symmetric, because the method targets skewness in the error term. Mathematically, if $\lambda < 1$, then the Box–Cox transformation pulls the right tail in more than it does the left tail, thus making right-skewed data more symmetric. If $\lambda > 1$, then the transformation pushes out the right tail more than the left tail. Therefore, when $\lambda < 1$, the Box–Cox transformation makes right-skewed data more symmetric; when $\lambda > 1$, it makes left-skewed data more symmetric. However, this transformation does not necessarily eliminate heavy tails.

Abrevaya (2002) provides the general theory of retransformation of the Box–Cox model under homoskedasticity. Duan's (1983) smearing for the lognormal model is a special case of Abrevaya's method.

In health economics, the Box–Cox transformation most commonly transforms a skewed dependent variable, such as positive expenditures. The log transformation is most common. The square root transformation has also been used in a few applications (Ettner et al. 1998, 2003; Lindrooth, Norton, and Dickey 2002; Veazie, Manning, and Kane 2003).

6.5.1 Box–Cox example

We next show one example of the basic Box–Cox transformation (dependent variable only) on the total expenditures variable from the MEPS data. The result is that $\widehat{\lambda} =$

0.052 (Stata calls this /theta). Although statistically significantly different from zero, this estimated transformation parameter is fairly close to zero—and justifies the use of the log model as the best simple approximation. If researchers were to encounter a case that was neither log nor linear, they could use the Abrevaya (2002) approach.

```
. *** Box-Cox test for total expenditures
. boxcox exp_tot if exp_tot > 0, nolog
Fitting comparison model

Fitting full model

                                              Number of obs   =    15,946
                                              LR chi2(0)      =      0.00
Log likelihood = -145803.04                   Prob > chi2     =         .
```

exp_tot	Coef.	Std. Err.	z	P>|z|	[95% Conf. Interval]
/theta	.0518686	.004114	12.61	0.000	.0438053 .0599318

Estimates of scale-variant parameters

	Coef.
Notrans	
_cons	8.88003
/sigma	2.375656

Test HO:	Restricted log likelihood	LR statistic chi2	P-value Prob > chi2
theta = -1	-181635.35	71664.62	0.000
theta = 0	-145882.99	159.91	0.000
theta = 1	-170429.36	49252.65	0.000

6.6 Stata resources

Models with log-transformed dependent variables are estimated with OLS regression, typically with **regress**. Generate a new logged dependent variable with **generate** prior to fitting the model.

The **boxcox** command estimates maximum likelihood estimates of Box–Cox models. There are four versions of the **boxcox** command: it will fit models with just the left-hand side transformed, models with just the right-hand side transformed, and models with both sides transformed—where the left- and right-hand sides have the same or different transformation factor. Stata's **predict** command implements the Abrevaya method after running **boxcox**. Therefore, **margins** gives consistent estimates of the means of the partial effects and standard errors.

Two commands related to **boxcox** will generate a new variable transformed from an old variable, such that the new variable has zero skewness. **bcskew0** uses the standard

Box–Cox formula to find a transformation with zero skewness. However, it differs slightly in its estimated parameter from `boxcox`, because it uses a different algorithm. `lnskew0` takes the natural logarithm of an expression minus a constant (k). Unlike `bcskew0`, `lnskew0` assumes a transformation parameter $(\lambda = 0)$, and chooses a value of the additive constant, k, to minimize the skewness of the transformed logged variable.

7 Models for continuous outcomes with mass at zero

7.1 Introduction

We have now explained a variety of ways to model positive outcomes with skewed positive values. While some important research questions in health economics and health policy involve only expenditures for those who spend at least some money, many more research questions involve health expenditures that include a substantial fraction of zeros—with the remaining values being positive, continuous, and severely skewed. For example, annual hospital expenditures are zero for most people but positive and often large for the subset who require hospital care. The majority of adults are nonsmokers, with many moderate smokers and a few heavy smokers. For any measure of healthcare use—inpatient, outpatient, emergency room, dental visit, preventive care—there is always a sizable fraction of the general population who do not use any healthcare during a defined period. The domains of all of these healthcare outcomes are either zero or positive. Statistical models that reflect the point mass at zero may better describe the relationships between the explanatory variables and the outcomes.

While it is tempting to eliminate zeros from the distributions of observed expenditures for statistical modeling reasons, incorporating them into the analysis is important for computing the correct treatment effects and for marginal and incremental effects of covariates. We often care about how treatments, policies, or other covariates affect the outcome for the entire population, including those who have zero expenditure or use. Some research questions are about whether a policy affects if a person has any expenditures (the extensive margin) or whether the policy affects the amount spent for those who have at least some expenditures (the intensive margin). An antismoking policy may affect the extensive margin (the fraction who smoke), the intensive margin (the number of cigarettes smoked by smokers), or both. Obtaining health insurance may also affect the intensive and extensive margins of healthcare spending differently. The most commonly used models that account for a substantial fraction of zeros allow for a differential response of the covariates over these two margins.

There are several ways to model such data, a number of which are discussed in Cameron and Trivedi (2005) and in Wooldridge (2010). In this chapter, we discuss two approaches in detail. Both approaches model the outcome using two indices; in each model, one index focuses on the process by which the zeros are generated. At the end of the chapter, we provide brief descriptions of single-index models that have been used in the literature but that we would not recommend.

We assume that the goal of the econometric strategy is to estimate conditional means [that is, $E(y_i|\mathbf{x}_i)$] and marginal and incremental effects of actual outcomes, [that is, $\partial E(y_i|\mathbf{x}_i)/\partial x$ and $E(y_i|\mathbf{x}_i, x = 1) - E(y_i|\mathbf{x}_i, x = 0)$], where y_i is the outcome for observation i, $\mathbf{x}_i$ is the vector of conditioning covariates, and x is a specific covariate. In most applications to health expenditures, researchers are not interested in the conditional expectation, $E(y_i^*|\mathbf{x}_i)$, of some underlying latent variable, y_i^*, in a model in which $y = 0$ denotes censoring—but instead $E(y_i|\mathbf{x}_i)$ in a model in which $y = 0$ truly represents $y = 0$. We compare these different statistical approaches to modeling continuous dependent variables with a large mass at zero, specifically on how they achieve the goals of predicting conditional means, marginal effects, and incremental effects of actual outcomes.

7.2   Two-part models

One approach to achieve these goals is based purely on a statistical decomposition of the data density (Cragg 1971). In this approach, it is assumed that the density of the outcome is a mixture of a process that generates zeros and a process that generates only positive values (this second process may not admit zeros). Consider an observed outcome, y_i, and a vector of covariates, $\mathbf{x}_i$.

Let f_0 be the density of y_i when $y_i = 0$, and let f_+ be the conditional density of y_i when $y_i > 0$. Without any loss of generality, we can write the density $g_i()$ of y_i as

$$g_i(y_i|\mathbf{x}_i) = \begin{cases} \{1 - \Pr(y_i > 0|\mathbf{x}_i)\} \times f_0(0|y_i = 0, \mathbf{x}_i) & \text{if } y_i = 0 \\ \Pr(y_i > 0|\mathbf{x}_i) \times f_+(y_i|y_i > 0, \mathbf{x}_i) & \text{if } y_i > 0 \end{cases} \qquad (7.1)$$

where $f_0(0|y_i = 0, \mathbf{x}_i) = 1$, because it defines a degenerate density at $y = 0$. By the fundamental definitions of joint and conditional events, the joint density $g_i(y_i|\mathbf{x}_i)$ always decomposes into the product of the probability that y_i is in a particular subdomain multiplied by its density, conditional on being in that subdomain. This definition is completely general. It does not require or imply any particular relationship between $\Pr()$ and f_+ (and f_0 to be precise). Specifically, we note that there is no independence requirement between the distributions or the stochastic elements that underly the distributions. Gilleskie and Mroz (2004) and Mroz (2012) invoke the same argument for a multiple index decomposition of a multivalued or count outcome. Drukker (2017) formally demonstrates you can identify $E(y_i|\mathbf{x}_i)$ when there is dependence between the part that determines whether $y = 0$ or $y > 0$ and the part that models $E(y_i|\mathbf{x}_i, y > 0)$.

The estimator of the parameters of this model can be decomposed into two parts; the parameters of the model for $\Pr(y_i > 0|\mathbf{x}_i)$ are estimated separately from the parameters of the model for $f_+(y_i|y_i > 0, \mathbf{x}_i)$. Because of this decomposition, this approach is widely known as the two-part model.

The two-part model has a long history in empirical analysis. Since the 1970s, meteorologists have used versions of a two-part model for rainfall (Cole and Sherriff 1972; Todorovic and Woolhiser 1975; Katz 1977). Economists also used two-part models in the 1970s. Cragg (1971) developed the hurdle (two-part) model as an extension of the tobit model. Newhouse and Phelps (1976) published an article that is the first known example of the two-part model in health economics. Their empirical model fits price and income elasticities of medical care. The two-part model became widely used in health economics and health services research after a team at the RAND Corporation used it to model healthcare expenditures in the context of the Health Insurance Experiment (Duan et al. 1984). See Mihaylova et al. (2011) for more on the widespread use of the two-part model for healthcare expenditure data. Two-part models are also appropriate for other mixed discrete-continuous outcomes, such as household-level consumption.

There are many specific modeling choices for the first- and second-part models. The choices depend on the data studied, the distribution of the outcome, and other statistical issues. The most common choices are displayed in table 7.1. In the first-part model, $\Pr(y_i > 0|\mathbf{x}_i)$ is typically specified as a logit or probit equation. In the second-part model, there are many suitable models for $E(y_i|y_i > 0, \mathbf{x}_i)$. Common choices are a linear model, a log-linear model (see chapter 6), or a generalized linear model (GLM) (see chapter 5).

Table 7.1. Choices of two-part models

First-part models	Comments
Probit	Often used, related to selection models
Logit	Common alternative to probit
Complementary log-log	Asymmetric p.d.f., used rarely

Second-part models	Comments
GLM	See chapter 5
Gaussian nonlinear least squares	GLM with $\delta = 0$
Poisson	GLM with $\delta = 1$
Gamma	GLM with $\delta = 2$
Wald, inverse Gaussian	GLM with $\delta = 3$
Box–Cox	See chapter 6
$\ln(y)$	Box–Cox with $\lambda = 0$
$\sqrt{y}$	Box–Cox with $\lambda = 0.5$
y	Box–Cox with $\lambda = 1$

7.2.1   Expected values and marginal and incremental effects

In this section, we describe how to compute the expected value of y, conditional on the vector of covariates $\mathbf{x}$, for different specific choices of the two parts of a two-part model. We also explain how to compute marginal and incremental effects. We focus on a few of the most popular two-part models, because there is not space to show every possible combination. By explaining the approach to the modeling and showing representative models, we leave it to readers to apply the models most appropriate to their own data.

Consider first a model with a probit first part and a normally distributed second part for a positive outcome, y, and vector of covariates, $\mathbf{x}$. The density, $g(y|\mathbf{x})$, is composed of two parts—depending on the value of y,

$$
g_i(y_i|\mathbf{x}_i) = \begin{cases} \{1 - \Phi(\mathbf{x}_i'\boldsymbol{\alpha}^p)\} & \text{if } y_i = 0 \\ \Phi(\mathbf{x}_i'\boldsymbol{\alpha}^p) \times \omega\phi\left(\frac{y_i - \mathbf{x}_i'\boldsymbol{\beta}}{\omega}\right) & \text{if } y_i > 0 \end{cases}
$$

where ϕ and Φ denote, respectively, the probability density function (p.d.f.) and the cumulative distribution function (CDF) of the unit normal density, $\boldsymbol{\alpha}^p$ is the vector of parameters for the first-part probit model, $\boldsymbol{\beta}$ is the vector of parameters for the second-part model, and ω is the scale (standard deviation) of the normal distribution in the second part. This model specification is even more restrictive than the usual linear second-part model which—if estimated by least squares—would not require normality. We use this restricted specification to aid comparison with the generalized tobit described in section 7.3.

We conclude this section by showing example formulas for the unconditional expected value of y_i, $E(y_i|\mathbf{x}_i)$. Because there are many different possible specifications for the two-part model, the formula for the unconditional expected value depends on the choice of models. For example, if the first part is probit and the second part is linear, then

$$
\begin{aligned}
E\left(y_i|y_i > 0, \mathbf{x}_i\right) &= \mathbf{x}_i'\boldsymbol{\beta} \\
E\left(y_i|\mathbf{x}_i\right) &= \Phi(\mathbf{x}_i'\boldsymbol{\alpha}^p) \times \mathbf{x}_i'\boldsymbol{\beta}
\end{aligned}
$$

If the first part is a probit and the second part is a GLM model with a log link, then the formula requires exponentiating the linear index function, where the vector of parameters is now denoted $\boldsymbol{\beta}^{\mathrm{GLM}}$:

$$
E\left(y_i|\mathbf{x}_i\right) = \Phi\left(\mathbf{x}_i'\boldsymbol{\alpha}^p\right) \times e^{\mathbf{x}_i'\boldsymbol{\beta}^{\mathrm{GLM}}}
$$

If instead the first part is a logit, then the first term on the right-hand side, $\Phi(\mathbf{x}_i'\boldsymbol{\alpha}^p)$, is replaced by the logit CDF $\{1 + \exp(-\mathbf{x}_i'\boldsymbol{\alpha}^l)\}^{-1}$, with a vector of parameters denoted $\boldsymbol{\alpha}^l$. For example, the two-part model with a logit and a GLM with a log link has an expected value of

$$
E\left(y_i|\mathbf{x}_i\right) = \frac{1}{\left(1 + e^{-\mathbf{x}_i'\boldsymbol{\alpha}^l}\right)} \times e^{\mathbf{x}_i'\boldsymbol{\beta}^{\mathrm{GLM}}}
$$

More work is necessary when the second part is ordinary least squares (OLS), with $\ln(y)$ as the dependent variable (see chapter 6). For example, if the first part is a probit, and the second part is a log transformation with homoskedastic normal errors, then

$$E(y_i|\mathbf{x}_i) = \Phi(\mathbf{x}_i'\boldsymbol{\alpha}^p) \times e^{\mathbf{x}_i'\boldsymbol{\beta}^{\ln y}} \times e^{0.5\tau^2}$$

where Φ is the normal CDF and τ^2 is the variance of the normal error ν. If the error ν is not assumed normal, then the term $\exp(0.5\tau^2)$ can be replaced by Duan's (1983) smearing factor, which we denote by D:

$$E(y_i|\mathbf{x}_i) = \Phi(\mathbf{x}_i'\boldsymbol{\alpha}^p) \times e^{\mathbf{x}_i'\boldsymbol{\beta}^{\ln y}} \times D$$

Other models require other formulas, but the expected value can always be calculated using the conditioning in (7.1).

7.3   Generalized tobit

The other approach is the generalized tobit (or Heckman) selection model, which begins with structural or behavioral equations that jointly model two latent outcomes. Each latent variable has an observed counterpart. Although we have formulated the model so that the outcome variable includes zeros and positives, following Maddala (1985), we note that the model was initially formulated as a combination of missing values and positives (Heckman 1979).

The generalized tobit explicitly models the correlation of the error terms of two structural equations, one for the censoring process and the other for the latent outcome. Using the notation of Wooldridge (2010), consider two latent random variables, w^* and y^*, with observed counterparts, w and y, respectively. The variable w^* defines a censoring process as

$$w_i = \begin{cases} 1 & \text{if } w_i^* > 0 \\ 0 & \text{if } w_i^* \le 0 \end{cases} \tag{7.2}$$

and an outcome equation as

$$y_i = \begin{cases} y_i^* & \text{if } w_i^* > 0 \\ 0 & \text{if } w_i^* \le 0 \end{cases} \tag{7.3}$$

Note that y is the observed outcome (for example, healthcare expenditures), with a mass of observations at zero. The model is completed by specifying the joint distribution of the latent variables, w^* and y^*. In this case,

$$\begin{aligned} w_i^* &= \mathbf{z}_i'\boldsymbol{\gamma} + \varepsilon_1 \\ y_i^* &= \mathbf{x}_i'\boldsymbol{\delta} + \varepsilon_2 \end{aligned} \tag{7.4}$$

where the vector, $\mathbf{z}$, is a superset of $\mathbf{x}$ (that is, $\mathbf{z}$ may include some variables not included in $\mathbf{x}$) and $\boldsymbol{\delta}$ and $\boldsymbol{\gamma}$ are vectors of parameters to estimate. If the joint distribution of ε_1 and ε_2 is bivariate normal with a correlation parameter, ρ,

$$\begin{bmatrix} \varepsilon_1 \\ \varepsilon_2 \end{bmatrix} \sim N(0, \boldsymbol{\Sigma}) \text{ and } \boldsymbol{\Sigma} = \begin{bmatrix} 1 & \rho\sigma \\ \rho\sigma & \sigma^2 \end{bmatrix} \tag{7.5}$$

Here $E(y_i^*|\mathbf{x}_i) = \mathbf{x}_i'\boldsymbol{\delta}$ and

$$\begin{aligned} E\left(y_i|y_i > 0, \mathbf{x}_i\right) &= \mathbf{x}_i'\boldsymbol{\delta} + \rho\sigma\lambda \\ E\left(y_i|\mathbf{x}_i\right) &= \Phi(\mathbf{z}_i'\boldsymbol{\gamma}) \times \{\mathbf{x}_i'\boldsymbol{\delta} + \rho\sigma\lambda\left(\mathbf{z}_i'\boldsymbol{\gamma}\right)\} \end{aligned} \tag{7.6}$$

where $\lambda(\mathbf{z}_i'\boldsymbol{\gamma}) = \{\phi(\mathbf{z}_i'\boldsymbol{\gamma})\}/\{\Phi(\mathbf{z}_i'\boldsymbol{\gamma})\}$ is the inverse Mills ratio.

The correlation, ρ, is identified from two sources. The preferred approach is to use exclusion restrictions. However, the model is also identified through nonlinearities in the functional form. In health economics applications, there is rarely a good justification for exclusion restrictions. Therefore, in practice, $\mathbf{z} = \mathbf{x}$ and identification is based entirely on functional form.

7.3.1 Full-information maximum likelihood and limited-information maximum likelihood

There are two standard ways to fit the selection model. The full selection model can be fit by full-information maximum likelihood (FIML). The likelihood function has one term for the probability that the main dependent variable is not observed, one term for the probability that it is observed (this term accounts for the error correlation), and one term for the positive conditional outcome assuming a normal error. If w is an indicator variable for whether y is observed, then the likelihood function is

$$L = \prod_{i=1}^{N} \{\Phi\left(-\mathbf{z}_i'\boldsymbol{\gamma}\right)\}^{1-w_i} \left\{ \Phi\left(\frac{\mathbf{z}_i'\boldsymbol{\gamma} + \{\ln\left(y_i\right) - \mathbf{x}_i'\boldsymbol{\delta}\} \frac{\rho}{\sigma}}{\sqrt{1-\rho^2}} \right) \right\}^{w_i}$$

$$\left\{ \frac{1}{\sigma\sqrt{2\pi}} e^{\frac{-1}{2}\left(\frac{\ln(y_i)-\mathbf{x}_i'\boldsymbol{\delta}}{\sigma}\right)^2} \right\}^{w_i}$$

Heckman (1979) proposed a computationally simpler limited-information maximum likelihood (LIML) estimator. Using LIML, you can fit the model in two steps—not to be confused with having two parts. The two steps of the LIML model can be fit sequentially. First, fit a probit model on the full sample of whether the outcome y is observed. Second, calculate the inverse Mills ratio, $\lambda = \phi/\Phi$, which is the ratio of the normal p.d.f. to the normal CDF. Finally, add the estimated inverse Mills ratio, $\widehat{\lambda}$, as a covariate to the main equation, and run OLS. The main equation is now

$$\begin{aligned} \ln\left(y_i|y_i > 0, \mathbf{x}_i\right) &= \mathbf{x}_i'\boldsymbol{\delta} + \rho\sigma\widehat{\lambda} + \nu \\ \widehat{\lambda} &= \left(\frac{\phi\left(\mathbf{z}_i'\widehat{\boldsymbol{\gamma}}\right)}{\Phi\left(\mathbf{z}_i'\widehat{\boldsymbol{\gamma}}\right)} \right) \end{aligned}$$

If $\rho = 0$, then the inverse Mills ratio drops out of the main equation, and the formula simplifies to a model without selection. There are several different definitions of the inverse Mills ratio, leading to different formulas that are close enough to be confusing. See the Stata FAQ for more discussion of why seemingly different formulas are actually equivalent.

Given that both the LIML and FIML estimators are consistent (under the usual assumptions), the choice between them falls to other considerations. Although both versions estimate ρ, FIML does it directly, while LIML estimates the combined parameter $(\rho\sigma)$—and ρ can be deduced given an estimate of σ. FIML sometimes fails to converge (especially if identification is only through nonlinear functional form), whereas LIML will always estimate its parameters. In Stata, LIML has a more limited set of postestimation commands, making it harder to compare with other models.

7.4 Comparison of two-part and generalized tobit models

The two-part and generalized tobit models look similar in many ways, but they have important differences, strengths, and weaknesses (Leung and Yu 1996; Manning, Duan, and Rogers 1987). It is therefore important to explain the fundamental differences between these models. There is a long-running debate in the health economics literature about the merits of the two-part model compared with the selection model (see Jones [2000] in the *Handbook of Health Economics* for a summary of the "cake debates"). The name "cake debates" comes from the title of one of the original articles comparing these models (Hay and Olsen 1984). Without delving into culinary metaphors or arbitrating the past debate directly, we make several points that focus on the salient statistical features that distinguish these two models.

First, the generalized tobit and two-part models are generally not nested models when each is specified parametrically. The many distinct versions of the two-part model make different assumptions about the first and second parts of the model. Most versions of the two-part model are not nested within the generalized tobit model.

Second, the generalized tobit is more general than one specific version of the two-part model. The generalized tobit, (7.2)–(7.5), with $\rho = 0$ and $\mathbf{z} = \mathbf{x}$, is formally equivalent to a two-part model with a probit first part and normally distributed second part. The generalized tobit with $\rho \neq 0$ is formally a generalization of this specific and restrictive version of the two-part model but is not a generalization of any other version of the two-part model.

Third, even for this case where the generalized tobit model is more general than the two-part model (a probit first part and a normally distributed second part), simulation evidence shows that the two-part model delivers virtually identical average marginal effects, the goal of our econometric investigation. More generally, Drukker (2017) formally demonstrates the equivalence of $E(y_i|\mathbf{x}_i)$ even if there is dependence in the data generating process. Nevertheless, this point is important enough that we will illustrate it with two examples—one with identification solely through functional form and one with an identifying excluded variable—in section 7.4.1.

Fourth, the two-part model can be motivated as a mixture density, which is at least as natural as a candidate data-generating process as that implied by the generalized tobit. Thus there is no compelling reason to view the two-part model as a special case of the generalized tobit; it can be motivated with a perfectly natural data-generating process that will not be nested within any generalized tobit model. For more on mixture densities, see chapter 9.

Fifth, the two-part model has an important practical advantage over the generalized tobit model. In the two-part model, it is trivially easy to change the specifications of both the first and second parts to allow for various error distributions and nonlinear functional forms (for example, logit or complementary log-log first parts and, more importantly, GLM or Box–Cox second parts). The different second-part models, discussed at length in chapters 5 and 6, are often important for dealing with statistical issues like skewness and heteroskedasticity on the positive values. Such changes require complex modifications in the generalized tobit, often leading to models that are not straightforward to estimate. Thus they are rarely implemented in practice.

Sixth, the standard interpretation of the models is different because of the original motivation for how to treat the zeros. The generalized tobit was originally intended to deal with missing values of the dependent variable, so it treats observed zeros as missing. For health economics, the standard interpretation of the generalized tobit model would therefore be to estimate what patients would pay if they had spent money. We are not aware of any articles that have been motivated by such a research question. The two-part model assumes that zeros are zeros (not missing values).

In conclusion, it is best to think of the two-part model (in all of its possible forms) and the generalized tobit as nonnested models. The point to note is that, in general, $E(y_i|\mathbf{x}_i)$ in the two-part model generally depends on how the functions and distributions in (7.1) are specified for the two-part model—whereas the form of $E(y_i|\mathbf{x}_i)$ in the context of the generalized tobit depends on how (7.6) and the associated joint distribution of the errors in those equations are specified. However, if interest is in the latent (uncensored) process a generalized tobit-type structure is essential. In that context, the parameter ρ plays a substantive role in interpretation of the parameters (Maddala 1983). Otherwise, especially if one is interested in understanding the conditional mean or marginal effects of covariates on that mean, the two-part model has greater practical appeal.

7.4.1   Examples that show similarity of marginal effects

The fact that the two-part model returns predictions and marginal effects that are virtually identical to those of a generalized tobit model—even when the data-generating process is for a generalized tobit—is so important and misunderstood that we present two illustrative examples. Drukker (2017) formally demonstrates this. In the first example, the data are generated using a generalized tobit data-generating process with jointly normal errors. There is no exclusion restriction ($\mathbf{z} = \mathbf{x}$), as is typical in health economics applications. Without loss of generality, the variance of the error term for the latent outcome is set equal to one.

```
. *** Example data-generating process identified only from functional form (z = x)
. set obs 10000
number of observations (_N) was 0, now 10,000
. set seed 123456
. matrix C1 = (1, .7, 1)                              // covariance matrix
. drawnorm eps1 eta1, cov(C1) cstorage(lower) double // draw errors
. generate double x1 = 1 + rnormal()                 // model for x1
. generate s1 = (-1 + x1 + eps1 >0)                  // model for s1
. generate double y1 = s1*(100 + x1 + eta1)          // model for y1
```

Comparing the estimated two-part (`twopm` in Stata) models and generalized tobit (`heckman`) models shows that the estimated coefficients are similar in the first equation (first-part and selection equations) but quite different in the second equation (second-part and main equations). In the second equations, the parameters on `x1` are 0.692 and 0.988. Although researchers might be tempted to conclude that these results imply that the models will lead to vastly different predictions of marginal effects, the marginal effects are in fact nearly identical—as seen in the Stata output:

```
. *** Results for two-part model and generalized tobit
. quietly twopm y1 x1, firstpart(probit) secondpart(regress)
. estimates store tpm1
. quietly heckman y1 x1, select(s1 = x1)
. estimates store heckman1
. estimates table tpm1 heckman1, se equations(1:2,2:1)
```

Variable	tpm1	heckman1
#1		
x1	.99578896	.99447454
	.01913407	.01888444
_cons	-1.026171	-1.0249822
	.02420804	.0240128
#2		
x1	.69174163	.98799387
	.01508252	.02731489
_cons	100.89028	100.06341
	.02687043	.06726254
/athrho		.76417581
		.06246671
/lnsigma		-.02512162
		.0182242

legend: b/se

Although parameter estimates of the second part of the two-part model do not correspond to those of the generalized tobit data-generating process, the marginal effect of x1 on y from the two-part model is virtually identical to those obtained from the generalized tobit model, 28.9:

```
. *** Marginal effects are nearly identical
. estimates restore tpm1
(results tpm1 are active now)
. quietly margins, dydx(x1) post
. estimates store tpmmargins1
. estimates restore heckman1
(results heckman1 are active now)
. quietly margins, dydx(x1) predict(yexpected) post
. estimates store heckmanmargins1
. estimates table tpmmargins1 heckmanmargins1, se
```

Variable	tpmmargi~1	heckmanm~1
x1	28.917742	28.899052
	.27617591	.27339699

legend: b/se

The second example has an identifying instrumental variable that can be excluded from the main equation. The data-generating process allows for a substantial effect of an additional variable, z, in the selection equation that does not enter the latent outcome equation. When the selection equation in the generalized tobit (7.2)–(7.5) data-generating process includes an excluded instrument—even if $\rho = 0$—the typical implementation of the two-part model would be overspecified, because it would include the same set of variables in both the first and second parts. Nevertheless, the simulation evidence shown in this example again highlights the flexibility of the two-part model specification.

```
. *** Example data-generating process with exclusion restriction z
. clear all
. set obs 10000
number of observations (_N) was 0, now 10,000
. set seed 123456
. matrix C2 = (1, .7, 1)                              // covariance matrix
. drawnorm eps2 eta2, cov(C2) cstorage(lower) double // draw errors
. generate double x2 = 1 + rnormal()                 // model for x
. generate double z2 = rnormal()                      // model for z
. generate s2 = (-1 + x2 + z2 + eps2 >0)             // model for s
. generate double y2 = s2*(100 + x2 + eta2)          // model for y
```

Again, the estimated coefficients in the two models are similar in the first equation but different in the second equation (0.821 versus 0.959).

```
. *** Results for two-part model and generalized tobit
. quietly twopm y2 x2, firstpart(probit) secondpart(regress)
. estimates store tpm2
. quietly heckman y2 x2, select(s2 = x2)
. estimates store heckman2
. estimates table tpm2 heckman2, se equations(1:2,2:1)
```

Variable	tpm2	heckman2
#1		
x2	.72387622	.7240433
	.01607619	.01606108
_cons	-.72541005	-.72555428
	.02108863	.02107642
#2		
x2	.82140422	.95931512
	.01475385	.04565497
_cons	100.59622	100.16343
	.02538826	.13743023
/athrho		.3876152
		.12321704
/lnsigma		-.03749737
		.02503957

legend: b/se

Despite the differences in estimated coefficients, the marginal effect of x2 on y from the two-part model is again virtually identical to that obtained from the generalized tobit model, 24.12. We care primarily about the estimates of the marginal effects on the expected outcomes, not the parameter estimates themselves.

```
. *** Marginal effects are nearly identical
. estimates restore tpm2
(results tpm2 are active now)
. *** Marginal effect for two-part model
. margins, dydx(x2)
```

Average marginal effects   Number of obs   =   10,000

Expression   : twopm combined expected values, predict()
dy/dx w.r.t. : x2

	dy/dx	Delta-method Std. Err.	z	P>|z|	[95% Conf. Interval]	
x2	24.12076	.3453853	69.84	0.000	23.44381	24.7977

```
. estimates restore heckman2
(results heckman2 are active now)
```

```
. *** Marginal effect for generalized tobit
. margins, dydx(x2) predict(yexpected)
Average marginal effects                    Number of obs      =      10,000
Model VCE     : OIM

Expression    : E(y2*|Pr(s2)), predict(yexpected)
dy/dx w.r.t.  : x2
```

	dy/dx	Delta-method Std. Err.	z	P>|z|	[95% Conf. Interval]	
x2	24.12441	.3450821	69.91	0.000	23.44806	24.80076

To summarize the third point, we see this simulation demonstrates that despite the apparent differences in model assumptions, the two-part model and the generalized tobit model usually produce similar results when comparing marginal effects of actual outcomes, which are usually the goal of econometric modeling in health economics. Now we return to the two-part model for interpretation and marginal effects.

7.5   Interpretation and marginal effects

7.5.1   Two-part model example

In this example, we use the 2004 Medical Expenditure Panel Survey (MEPS) data introduced in chapter 3 to estimate the effect of age (age) and gender (female is a binary indicator for being female) on total healthcare expenditures (exp_tot). In this two-part model, we use a probit model to predict the probability of any expenditures and a GLM model with a log link and gamma family to predict the level of expenditures for those who have more than zero. The goals are to estimate total expenditures conditional on the covariates and then to calculate the marginal effect of age and the incremental effect of gender. To focus on technique, we limit the covariates to just age and gender and their interaction.

The results below could be computed separately, first by fitting two models (probit and then either GLM or OLS), but instead we use the twopm (two-part model) Stata command written by Belotti et al. (2015). This allows for easier computation of predicted values and marginal effects using the postestimation commands.

In both parts, the estimated coefficients for age and female are positive and statistically significant at the one-percent level, while the interaction term is negative and statistically significant. Both the probability of spending and the amount of spending conditional on any spending increase with age but at a slower rate for women. Women are more likely to spend at least $1 more than men, and, conditional on spending any amount, they spend more, at least at younger ages. The results for the second part of the model are the same as in the first simple GLM example in section 5.3.

```
. *** twopm: 1) probit  2) GLM with log link and gamma distribution
. twopm exp_tot c.age##i.female, firstpart(probit, nolog)
> secondpart(glm, family(gamma) link(log) nolog) vce(robust)
Fitting probit regression for first part:
Fitting glm regression for second part:
Two-part model
```

Log pseudolikelihood = -156026.29	Number of obs	= 19386

Part 1: probit

	Number of obs	= 19386
	Wald chi2(3)	= 1929.84
	Prob > chi2	= 0.0000
Log pseudolikelihood = -8042.5542	Pseudo R2	= 0.1126

Part 2: glm

		Number of obs	= 15946
Deviance	= 33350.09564	(1/df) Deviance =	2.091964
Pearson	= 120642.3512	(1/df) Pearson =	7.567579
Variance function: V(u) = u^2		[Gamma]	
Link function : g(u) = ln(u)		[Log]	
		AIC	= 18.56111
Log pseudolikelihood = -147983.7344		BIC	= -120920.1

exp_tot	Coef.	Robust Std. Err.	z	P>|z|	[95% Conf. Interval]	
probit						
age	.0303793	.0010029	30.29	0.000	.0284137	.032345
female						
Female	.8773617	.061366	14.30	0.000	.7570864	.9976369
female#c.age						
Female	-.0086631	.0014161	-6.12	0.000	-.0114386	-.0058877
_cons	-.5964252	.0433537	-13.76	0.000	-.6813969	-.5114536
glm						
age	.0345881	.0024311	14.23	0.000	.0298233	.0393529
female						
Female	.7164142	.1614244	4.44	0.000	.4000283	1.0328
female#c.age						
Female	-.0106117	.0027302	-3.89	0.000	-.0159628	-.0052607
_cons	6.513084	.1468077	44.36	0.000	6.225347	6.800822

```
. estimates store lntot_twopm
```

After we fit both parts of the two-part model with `twopm`, the postestimation `margins` command calculates predictions based on both parts. The predicted total spending is about \$3,696 per person per year, which is within a few dollars of the actual average (\$3,685).

```
. *** Overall average prediction of total expenditures
. margins
Warning: cannot perform check for estimable functions.
Predictive margins                          Number of obs   =     19,386
Model VCE    : Robust

Expression   : twopm combined expected values, predict()
```

	Margin	Delta-method Std. Err.	z	P>|z|	[95% Conf. Interval]
_cons	3695.711	68.74228	53.76	0.000	3560.979 3830.443

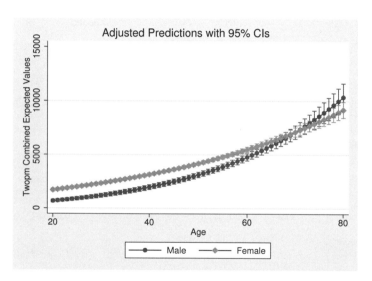

Figure 7.1. Predicted expenditures, conditional on age and gender

The predicted total expenditures, combining both parts of the two-part model, confirm what the coefficients implied. Expenditures are higher for women than for men at younger ages but rise faster for men, with the crossover point being around age 70 (see figure 7.1).

7.5.2   Two-part model marginal effects

This section outlines how to compute marginal and incremental effects in two-part models, accounting for the full model. We use simple notation throughout this section to show the structure of the formulas. Formula details would depend on which specific version of the two-part model is fit. The main formula is

$$E\left(y_i|\mathbf{x}_i\right) = \Pr\left(y_i > 0|\mathbf{x}_i\right) \times E\left(y_i|y_i > 0, \mathbf{x}_i\right) \tag{7.7}$$

The marginal effect of a continuous variable, x, on the expected value of y is the full derivative of (7.7). Therefore, the full marginal and incremental effects include both the extensive margin (effect on the probability that $y > 0$) and the intensive margin (effect on the mean of y conditional on $y > 0$). The marginal effect is computed by the chain rule:

$$\frac{\partial E\left(y_i|\mathbf{x}_i\right)}{\partial x} = \Pr\left(y_i > 0|\mathbf{x}_i\right) \times \frac{\partial E\left(y_i|y_i > 0, \mathbf{x}_i\right)}{\partial x}$$

$$+ \frac{\partial \Pr\left(y_i > 0|\mathbf{x}_i\right)}{\partial x} \times E\left(y_i|y_i > 0, \mathbf{x}_i\right)$$

For the case of a probit first-part model and a GLM second-part model (and no interactions or higher order terms in $\mathbf{x}$), this is fairly straightforward to compute,

$$\frac{\partial E\left(y_i|\mathbf{x}_i\right)}{\partial x} = \Phi\left(\mathbf{x}_i'\boldsymbol{\alpha}^p\right)\left(\beta_x^{\mathrm{GLM}} e^{\mathbf{x}_i'\boldsymbol{\beta}^{\mathrm{GLM}}}\right) + \left\{\alpha_x^p \phi\left(\mathbf{x}_i'\boldsymbol{\alpha}^p\right)\right\} e^{\mathbf{x}_i'\boldsymbol{\beta}^{\mathrm{GLM}}}$$

where $\boldsymbol{\alpha}^p$ is the vector of parameters in the first-stage probit, $\boldsymbol{\beta}^{\mathrm{GLM}}$ is the vector of parameters in the second-stage GLM, and α_x^p and β_x^{GLM} refer to the coefficients corresponding to the specific covariate, x.

If there were interactions with x or higher-order terms in x, then the derivatives would need to account for the expressions in brackets.

For OLS models with a log-transformed y (and corresponding vector of parameters $\boldsymbol{\beta}^{\ln y}$), the marginal effects depend on how you deal with heteroskedasticity. First, we show formulas assuming homoskedasticity. These calculations can be manipulated to be expressions of either $E(y|y > 0, \mathbf{x})$ or $E(y|\mathbf{x})$. For the probit model, this is

$$p_i = \Pr\left(y_i > 0|\mathbf{x}_i\right) = \Phi\left(\mathbf{x}_i'\boldsymbol{\alpha}^p\right)$$

$$\frac{\partial E\left(y_i|\mathbf{x}_i\right)}{\partial x} = (p_i)\left(\beta_x^{\ln y} e^{\mathbf{x}_i'\boldsymbol{\beta}^{\ln y}} D\right) + (\alpha_x^p \phi)\left(e^{\mathbf{x}_i'\boldsymbol{\beta}^{\ln y}} D\right)$$

$$= e^{\mathbf{x}_i'\boldsymbol{\beta}^{\ln y}} D\left(p_i \beta_x^{\ln y} + \alpha_x^p \varphi\right)$$

$$= E\left(y_i|y_i > 0, \mathbf{x}_i\right)\left(p_i \beta_x^{\ln y} + \alpha_x^p \varphi\right)$$

where D is Duan's smearing factor.

For the logit model, this is

$$p_i = \Pr\left(y_i > 0|\mathbf{x}_i\right) = \frac{1}{1 + e^{-\mathbf{x}_i'\boldsymbol{\alpha}^l}}$$

$$\frac{\partial E\left(y_i|\mathbf{x}_i\right)}{\partial x} = (p_i)\left(\beta_x^{\ln y} e^{\mathbf{x}_i'\boldsymbol{\beta}^{\ln y}} D\right) + \left\{\alpha_x^l p_i (1 - p_i)\right\}\left(e^{\mathbf{x}_i'\boldsymbol{\beta}^{\ln y}} D\right)$$

$$= p_i e^{\mathbf{x}_i'\boldsymbol{\beta}^{\ln y}} D\left\{\beta_x^{\ln y} + \alpha_x^l(1 - p_i)\right\}$$

$$= E\left(y_i|\mathbf{x}_i\right)\left\{\beta_x^{\ln y} + \alpha_x^l(1 - p_i)\right\}$$

In contrast to a marginal effect, where it makes sense to think of a tiny increase in the value of a continuous variable for a dummy variable, we compute an incremental effect.

Consider the dichotomous variable `female`. Conceptually, we compute the predicted value of the outcome two ways, first as if everyone in the sample is female, then as if everyone in the sample is male (always holding all other covariates at their original levels), and then take the difference. More generally, for a dichotomous indicator, x:

$$\frac{\Delta E\left(y_i|\mathbf{x}\right)}{\Delta x} = \{\Pr\left(y > 0|\mathbf{x}_i, x = 1\right) - \Pr\left(y_i > 0|\mathbf{x}_i, x = 0\right)\} \times E\left(y_i|y_i > 0, \mathbf{x}_i\right)$$

$$+ \Pr\left(y_i > 0|\mathbf{x}_i\right) \times \{E\left(y_i|y_i > 0, \mathbf{x}_i, x = 1\right) - E\left(y_i|y_i > 0, \mathbf{x}_i, x = 0\right)\}$$

$$\frac{\Delta E\left(y_i|\mathbf{x}\right)}{\Delta x} = \left(p_i^1 - p_i^0\right) E_i + p_i \left(E_i^1 - E_i^0\right)$$

If the second part of the model is heteroskedastic, the marginal and incremental effects are more complicated, because the smearing factor D is no longer a scalar. See Manning (1998) and Ai and Norton (2000, 2008) for methods to deal with heteroskedasticity when retransforming back to the raw scale.

7.5.3 Two-part model marginal effects example

Continuing the example from section 7.5.1, we now show the marginal (or incremental) effects of age and gender for the full two-part model, accounting for the effects of these variables on both parts. After we use the `twopm` command, the `margins` command automatically computes the unconditional marginal effects, accounting for both parts of the model. The marginal effect of age averages \$123 per year of age, and women spend more than men by about \$798.

```
. *** Marginal effects of age and gender
. margins, dydx(*)
Warning: cannot perform check for estimable functions.
Average marginal effects                          Number of obs   =     19,386
Model VCE      : Robust

Expression     : twopm combined expected values, predict()
dy/dx w.r.t.   : age 1.female
```

	dy/dx	Delta-method Std. Err.	z	P>|z|	[95% Conf. Interval]	
age	123.2596	4.932306	24.99	0.000	113.5924	132.9267
female						
Female	797.6211	142.5512	5.60	0.000	518.2258	1077.016

```
Note: dy/dx for factor levels is the discrete change from the base level.
```

Because the graphs showed that the marginal effects vary over the life course, we computed marginal effects, conditional on four ages (20, 40, 60, and 80). The marginal effect of age for men grows from \$40 at age 20 to \$383 by age 80; the marginal effect of age for women grows from \$56 at age 20 to \$231 by age 80. The incremental effect of gender declines, with women spending on average more than \$1,000 more than men at age 20, but by age 80, the roles have reversed, and men outspend women by more than $1,000.

```
. *** Marginal effect of age at different ages
. margins, dydx(*) at(female=(0 1) age=(20(20)80))
Warning: cannot perform check for estimable functions.
Conditional marginal effects                     Number of obs    =     19,386
Model VCE     : Robust

Expression    : twopm combined expected values, predict()
dy/dx w.r.t.  : age 1.female

1._at         : age             =           20
                female          =            0

2._at         : age             =           20
                female          =            1

3._at         : age             =           40
                female          =            0

4._at         : age             =           40
                female          =            1

5._at         : age             =           60
                female          =            0

6._at         : age             =           60
                female          =            1

7._at         : age             =           80
                female          =            0

8._at         : age             =           80
                female          =            1
```

		dy/dx	Delta-method Std. Err.	z	P>|z|	[95% Conf. Interval]	
age							
	_at						
	1	39.7952	2.530461	15.73	0.000	34.83559	44.75481
	2	55.70189	1.260417	44.19	0.000	53.23152	58.17226
	3	94.95937	3.474961	27.33	0.000	88.14857	101.7702
	4	91.60674	3.267144	28.04	0.000	85.20326	98.01023
	5	195.941	14.07623	13.92	0.000	168.3521	223.5299
	6	145.8835	8.860599	16.46	0.000	128.5171	163.25
	7	382.7012	47.00823	8.14	0.000	290.5667	474.8356
	8	230.8949	20.11701	11.48	0.000	191.4663	270.3235
0.female		(base outcome)					
1.female							
	_at						
	1	1020.784	103.6669	9.85	0.000	817.601	1223.968
	2	1020.784	103.6669	9.85	0.000	817.601	1223.968
	3	1181.387	138.0487	8.56	0.000	910.8166	1451.957
	4	1181.387	138.0487	8.56	0.000	910.8166	1451.957
	5	707.1014	210.9425	3.35	0.001	293.6616	1120.541
	6	707.1014	210.9425	3.35	0.001	293.6616	1120.541
	7	-1178.08	749.5238	-1.57	0.116	-2647.12	290.9596
	8	-1178.08	749.5238	-1.57	0.116	-2647.12	290.9596

Note: dy/dx for factor levels is the discrete change from the base level.

7.5.4   Generalized tobit interpretation

There are three standard ways to interpret the results from the generalized tobit model. The first way focuses on what happens to the expected latent outcome (denoted here y^*). Latent outcomes assume that the dependent variable is missing (not zero) for part of the sample and that the selection equation adequately adjusts for the nonrandom selection. The expected value of the latent outcome is $\mathbf{x}'\widehat{\boldsymbol{\delta}}$, using the same notation as in section 7.3. The first interpretation is easy to read from the regression output table but not relevant for answering research questions in health economics, where we typically care about predictions of actual expenditures.

The other two interpretations for the generalized tobit are more challenging to calculate. The second interpretation focuses on the characteristics of the actual outcome and is therefore comparable with the results from a two-part model (Duan et al. 1984; Poirier and Ruud 1981; Dow and Norton 2003). In this case,

$$E\left(y_i|\mathbf{x}_i\right) = \Phi\left(\mathbf{z}_i'\widehat{\boldsymbol{\gamma}}\right)\left(\mathbf{x}_i'\widehat{\boldsymbol{\delta}} + \widehat{\rho}\widehat{\sigma}\widehat{\lambda}\right)$$

The third interpretation is the expected value of the observed outcome conditional on observing the dependent variable and is

$$E\left(y_i|y_i > 0, \mathbf{x}_i\right) = \mathbf{x}_i'\widehat{\boldsymbol{\delta}} + \widehat{\rho}\widehat{\sigma}\widehat{\lambda}$$

Clearly, these three different expected values will differ in magnitude, as will the associated marginal effects. The marginal effect of a covariate on the latent outcome, y^*, is simply $\widehat{\gamma}_k$. However, the marginal effect of a covariate on the actual outcome, y, is complicated. If y is linear, with no logs or retranformation issues, then

$$\frac{\partial E\left(y_i|\mathbf{x}_i\right)}{\partial x_k} = \widehat{\gamma}_k\Phi\left(\mathbf{z}_i'\widehat{\boldsymbol{\gamma}}\right) + \widehat{\delta}_k\phi\left(\mathbf{z}_i'\widehat{\boldsymbol{\gamma}}\right)\left(\mathbf{x}_i'\widehat{\boldsymbol{\delta}} + \widehat{\rho}\widehat{\sigma}\mathbf{z}_i'\widehat{\boldsymbol{\gamma}}\right)$$

If instead the main outcome is estimated as $\ln\left(y_i|y_i > 0, \mathbf{x}_i\right) = \mathbf{x}_i'\boldsymbol{\delta} + \varepsilon_i$ and the error term is normal and homoskedastic, then

$$
\begin{aligned}
E\left(y_i|\mathbf{x}_i\right) &= \Phi\left(\mathbf{z}_i'\widehat{\boldsymbol{\gamma}} + \widehat{\rho}\widehat{\sigma}\right)\exp\left(\mathbf{x}_i'\widehat{\boldsymbol{\delta}} + .5\widehat{\sigma}^2\right) \\
\frac{\partial E\left(y_i|\mathbf{x}_i\right)}{\partial x_k} &= \left\{\widehat{\gamma}_k + \widehat{\delta}_k\widehat{\lambda}\left(\mathbf{z}_i'\widehat{\boldsymbol{\gamma}} + \widehat{\rho}\widehat{\sigma}\right)\right\} E\left(y_i|\mathbf{x}_i\right)
\end{aligned}
$$

Any other model specification would require different specific formulas.

7.5.5   Generalized tobit example

Although we wanted to directly compare the results from the two-part model on total healthcare expenditures with the results from the generalized tobit (or Heckman) selection model, it is not possible. For health expenditures in the MEPS data, FIML fails to converge. The FIML estimator is highly sensitive to departures from joint normality, and the positive values of total healthcare expenditures are not close to normal. Having a highly skewed distribution of the positive values is a common issue in health economics. Although we can fit the model with a logged dependent variable, it makes comparisons with the two-part model much harder, because expressing marginal effects on the raw scale cannot be done automatically in Stata. LIML estimates are easy to obtain, but the postestimation commands that compute marginal effects do not work in Stata for that case.

Therefore, to make comparisons easier across the two-part, FIML, and LIML models, we changed the example to analyze dental expenditures. We use the MEPS data introduced in chapter 3 to estimate the effect of age (`age`) and gender (`female` is a binary indicator of being female) on dental health expenditures, including the many people with zero expenditures.

```
. *** Dental expenditures compared with two-part model
. use http://www.stata-press.com/data/heus/heus_mepssample, clear
(Sample of MEPS 2004 data)
. generate y_heckman = exp_dent if exp_dent > 0
(12,147 missing values generated)
```

The results for the three models of dental expenditures have nearly identical coefficients in the first equation (probit), which is not surprising. The coefficients in the second equation are different, especially those from the two-part model, because there we have modeled the conditional mean to be an exponential function of the linear index. The coefficients obtained by FIML and LIML are also quite different from each other, partly because the Mills ratio is both large and imprecisely estimated. However, as the results below show, the marginal effects are similar across all three models.

```
. quietly twopm exp_dent c.age##i.female, firstpart(probit, nolog)
> secondpart(glm, family(gamma) link(log) nolog) vce(robust)
. estimates store lndent_twopm
. quietly heckman y_heckman c.age##i.female, select(c.age##i.female)
. estimates store dent_FIML
. quietly heckman y_heckman c.age##i.female, select(c.age##i.female) twostep
. estimates store dent_LIML
```

```
. estimates table lndent_twopm dent_FIML dent_LIML, se equations(1:2:2,2:1:1)
```

Variable	lndent_t~m	dent_FIML	dent_LIML
#1			
age	.00988265	.00988265	.00988265
	.00081856	.00082111	.00082111
female			
Male	(empty)	(base)	(base)
Female	.42408762	.42407677	.42408762
	.05213285	.05258829	.0525877
female#c.age			
Male	(empty)	(base)	(base)
Female	-.00597587	-.00597571	-.00597587
	.0010682	.0010749	.00107489
_cons	-.85711944	-.85711615	-.85711944
	.0395746	.03982663	.0398264
#2			
age	.00776295	4.3350718	-11.950836
	.00198979	1.523104	59.960118
female			
Male	(empty)	(base)	(base)
Female	.06866129	42.983943	-667.77652
	.12838514	84.237146	2618.5235
female#c.age			
Male	(empty)	(base)	(base)
Female	-.001045	-.73124353	9.330016
	.00251631	1.5378523	37.137797
_cons	5.9520068	366.76039	3610.7349
	.10179496	222.56904	11923.744
/athrho		-.01151777	
		.15867363	
/lnsigma		6.8829107	
		.00840869	
/mills			-2336.6066
			8543.8616

legend: b/se

We restore the two-part model results to use `margins`. Overall, average dental expenditures are \$211, according to the two-part model results.

```
. *** Predict dental expenditures from two-part model
. estimates restore lndent_twopm
(results lndent_twopm are active now)
. margins
Warning: cannot perform check for estimable functions.
Predictive margins                          Number of obs     =     19,386
Model VCE    : Robust
Expression   : twopm combined expected values, predict()
```

	Margin	Delta-method Std. Err.	z	P>|z|	[95% Conf. Interval]	
_cons	211.4595	4.715068	44.85	0.000	202.2182	220.7009

Women spend more than men on average over all ages by almost \$32. Dental expenditures increase on average by about \$2.87 per year.

```
. *** Predict marginal effects for dental expenditures from two-part model
. margins, dydx(*)
Warning: cannot perform check for estimable functions.
Average marginal effects                    Number of obs     =     19,386
Model VCE    : Robust
Expression   : twopm combined expected values, predict()
dy/dx w.r.t. : age 1.female
```

	dy/dx	Delta-method Std. Err.	z	P>|z|	[95% Conf. Interval]	
age	2.869215	.2899934	9.89	0.000	2.300838	3.437592
female Female	31.95432	9.439078	3.39	0.001	13.45406	50.45457

Note: dy/dx for factor levels is the discrete change from the base level.

The marginal effect of age is higher for men than for women at all ages.

```
. *** Dental marginal effects for two-part model
. estimates restore lndent_twopm
(results lndent_twopm are active now)

. margins, dydx(age) at(female=(0 1) age=(25 65))
Warning: cannot perform check for estimable functions.
Conditional marginal effects                     Number of obs    =      19,386
Model VCE    : Robust

Expression   : twopm combined expected values, predict()
dy/dx w.r.t. : age

1._at       : age            =           25
              female         =            0

2._at       : age            =           25
              female         =            1

3._at       : age            =           65
              female         =            0

4._at       : age            =           65
              female         =            1
```

		dy/dx	Delta-method Std. Err.	z	P>|z|	[95% Conf. Interval]	
age	_at						
	1	2.510089	.1818655	13.80	0.000	2.15364	2.866539
	2	1.924377	.2401017	8.01	0.000	1.453787	2.394968
	3	4.505511	.7289859	6.18	0.000	3.076725	5.934297
	4	2.814159	.53872	5.22	0.000	1.758288	3.870031

For comparison with the two-part model, we must use the formulas for actual expenditures with the FIML results. It is important to use the predict(yexpected) option to calculate predictions for actual expenditures, not latent expenditures—otherwise the results are not directly comparable. Again, predicted actual expenditures for the two-part model and generalized tobit are quite close, certainly well within confidence intervals, even with vastly different estimated coefficients.

```
. *** Predicted dental expenditures from FIML
. estimates restore dent_FIML
(results dent_FIML are active now)

. margins, predict(yexpected)
Predictive margins                               Number of obs    =      19,386
Model VCE    : OIM

Expression   : E(y_heckman*|Pr(select)), predict(yexpected)
```

	Margin	Delta-method Std. Err.	z	P>|z|	[95% Conf. Interval]	
_cons	211.2604	4.703602	44.91	0.000	202.0415	220.4793

```
. margins, dydx(*) predict(yexpected)
Average marginal effects                        Number of obs      =      19,386
Model VCE     : OIM
Expression    : E(y_heckman*|Pr(select)), predict(yexpected)
dy/dx w.r.t.  : age 1.female
```

	dy/dx	Delta-method Std. Err.	z	P>|z|	[95% Conf. Interval]	
age	2.83416	.2763821	10.25	0.000	2.292461	3.375859
female Female	32.00451	9.423246	3.40	0.001	13.53529	50.47374

```
Note: dy/dx for factor levels is the discrete change from the base level.
```

The FIML-estimated marginal effects are also quite similar to those for the two-part model.

```
. *** Dental marginal effects for FIML
. margins, dydx(age) at(female=(0 1) age=(25 65)) predict(yexpected)
Conditional marginal effects                    Number of obs      =      19,386
Model VCE     : OIM
Expression    : E(y_heckman*|Pr(select)), predict(yexpected)
dy/dx w.r.t.  : age
1._at         : age              =          25
                female           =           0
2._at         : age              =          25
                female           =           1
3._at         : age              =          65
                female           =           0
4._at         : age              =          65
                female           =           1
```

		dy/dx	Delta-method Std. Err.	z	P>|z|	[95% Conf. Interval]	
age							
_at							
	1	2.707209	.2375702	11.40	0.000	2.24158	3.172838
	2	2.059558	.3144066	6.55	0.000	1.443332	2.675783
	3	4.288344	.5739972	7.47	0.000	3.16333	5.413358
	4	2.530505	.4460158	5.67	0.000	1.656331	3.40468

In sharp contrast to the actual outcomes, the results from FIML can also be used to compute latent outcomes, which is the default Stata option. Because about 63% of the sample has zero dental expenditures, if instead they all spent an average amount, then the total would of course more than double. That is exactly what is shown.

```
. *** Latent outcomes for FIML
. estimates restore dent_FIML
(results dent_FIML are active now)

. margins

Predictive margins                          Number of obs     =      19,386
Model VCE     : OIM

Expression    : Linear prediction, predict()
```

	Margin	Delta-method Std. Err.	z	P>|z|	[95% Conf. Interval]	
_cons	568.55	158.0928	3.60	0.000	258.6937	878.4063

```
. margins, dydx(*)

Average marginal effects                    Number of obs     =      19,386
Model VCE     : OIM

Expression    : Linear prediction, predict()
dy/dx w.r.t.  : age 1.female
```

	dy/dx	Delta-method Std. Err.	z	P>|z|	[95% Conf. Interval]	
age	3.933201	.9888248	3.98	0.000	1.99514	5.871262
female Female	9.814093	28.94926	0.34	0.735	-46.92541	66.55359

```
Note: dy/dx for factor levels is the discrete change from the base level.
```

In summary, if you want to estimate actual outcomes and marginal effects on actual outcomes (as opposed to latent outcomes), the FIML selection model will typically yield similar results to the two-part model. However, in practice, researchers fit two-part models because the results are easier to manipulate, both for the total effect and for the extensive and intensive margins.

7.6   Single-index models that accommodate zeros

In this section, we briefly describe some single-index models that allow for a mass of zeros in the distribution of the outcome but not in particularly flexible ways. We describe these models because they have been used in the literature, but we cannot recommend their use in research.

7.6.1   The tobit model

The tobit model, named after economist James Tobin, is like a mermaid or centaur; it is half one thing and half another. Tobin (1958) was the first to model dependent variables with a large fraction of zeros. Specifically, the tobit model combines the probit model with OLS, both in the way the model is fit and in how it is interpreted. For a recent summary of the tobit model, see Enami and Mullahy (2009).

The classic tobit model is appropriate when the dependent variable has two properties:

- it has a normal distribution (this is a strong assumption); and
- negative values are censored at zero.

Censored means that the actual value is known to be beyond a threshold value, or less than zero in this case. Censoring is different from truncation, in which there is no information about the actual value—so it is missing. The classic tobit model is rarely, if ever, appropriate for modeling healthcare expenditures, because zero expenditures are not censored negative values—but instead are actual values.

The tobit model assumes that y^* is a continuous, semiobserved (censored), normally distributed, underlying latent dependent variable. Semiobserved means that some values of y^* are observed, and other values are known only to be in a range. The tobit model fits the relationship between covariates and the latent variable, y^*.

Specifically, the classic tobit model assumes that the latent variable, y^*, can be negative—but that when y^* is negative, the observed value, y, is zero.

$$
\begin{aligned}
y_i^* &= \mathbf{x}_i'\boldsymbol{\delta} + \varepsilon_i \\
y_i &= \begin{cases} 0 & \text{if} \quad y_i^* \leq 0 \\ y_i^* & \text{if} \quad y_i^* > 0 \end{cases}
\end{aligned}
$$

The values equal to zero are censored, because they are recoded from a true negative value to zero. (If instead those observations were left out of the sample, they would be truncated, which is a selection problem.)

The tobit likelihood function has two pieces. There is the probability that observed y equals zero, and the probability that y equals some positive value. If ε has a normal distribution with variance σ^2, and if w is an indicator variable for whether y is positive, then the likelihood function is written as part normal CDF and part normal p.d.f..

$$
L = \prod_{i=1}^{N} \Phi\left(\frac{-\mathbf{x}_i'\boldsymbol{\delta}}{\sigma}\right)^{1-w_i} \left\{\frac{1}{\sigma}\phi\left(\frac{y_i - \mathbf{x}_i'\boldsymbol{\delta}}{\sigma}\right)\right\}^{w_i}
$$

The likelihood function is written in terms of the error term, specifically, a standard normal error term (mean 0 and variance 1). The $1/\sigma$ in the p.d.f. is the Jacobian term, the result of a normalization needed when doing a linear transformation of ε to a uniform normal variable. The derivative of the normal CDF with respect to the error term is the normal p.d.f. multiplied by the inverse of the error's standard deviation:

$$
\frac{\partial \Phi\left(\frac{\varepsilon}{\sigma}\right)}{\partial \varepsilon} = \frac{1}{\sigma}\phi\left(\frac{\varepsilon}{\sigma}\right)
$$

The number of parameters is the same as OLS and one more than for probit. There is one set of $\boldsymbol{\delta}$'s (including the constant) and one σ. There is no estimated parameter for the censoring point (zero in this case), because this threshold is not estimated; it is determined by the data.

How does the tobit model differ from the probit model? The tobit model fits one more parameter than probit. The tobit model has a continuous part and a discrete part. The interpretation of the constant term is quite different—for tobit, it has the interpretation of an OLS intercept, and for probit, it has the interpretation of the probability of outcome A for the base case observation.

The tobit model is extremely sensitive to its underlying assumptions of normality and homoskedasticity (Hurd 1979; Goldberger 1981).

7.6.2   Why tobit is used sparingly

The tobit model should be used with great caution, if at all. The assumptions underlying the model are numerous and rarely true. The tobit model should never be fit unless the data are truly normal and censored. Here are the top four reasons to avoid the tobit model:

1. The tobit model assumes that the data are censored at zero, instead of actually being zero. Too often, researchers with health expenditure data claim that a large mass at zero are censored observations when they are not censored.

2. The tobit model assumes that the error term has a normal distribution but is inconsistent even if there are minor departures from normality and homoskedasticity (Hurd 1979; Goldberger 1981).

3. The tobit model assumes that the error term when y is positive is truncated normal, with the truncation point at zero. This is rarely true.

4. The tobit model assumes that the same parameters govern both parts of the likelihood function. There are specification tests that test the tobit model against the more general Cragg (1971) model that allows different parameters in the two parts of the model. This test almost always rejects the null hypothesis that the parameters in both parts are equal.

In summary, the classic tobit model only applies in the rare cases where zero values are truly censored. Right-censoring is more common in real data, and tobit models may work well in those cases.

The tobit model has been used only a few times in the health economics literature. Holmes and Deb (1998) use a tobit model for data on health expenditures that are right-censored. The dependent variable they study is health expenditures for an episode of care. Because they have claims data for a calendar year, some episodes of care are artificially censored at the end of December. Cook and Moore (1993) use a tobit to estimate drinks per week. However, there is no evidence that abstainers are appropriately modeled as censored.

7.6.3   One-part models

Although two-part models are popular they are not the only estimation approach for addressing a large mass at zero. Mullahy (1998) suggested that researchers not use two-part models—especially those that use the log transformation in the conditional part—if they are interested in the expected value of y given observed covariates, $\mathbf{x}$. Using nonlinear least squares, or some variant of the GLM family, researchers can apply a single model to all the data to fit the expected value of y. Any of the links and families described in chapter 5 could be used for a one-part model as an alternative to a two-part model, as long as researchers are interested in the mean function for y, conditional on the covariates $\mathbf{x}$—or something that can be derived from the mean function, such as the marginal or incremental effect.

Some analysts have worried that some of the distributions used in the GLM approach do not have zeros in their support. This is a problem if the models are fit by maximum likelihood estimation (MLE). However, the GLM approach only uses mean and variance functions. For example, for the inverse Gaussian (Wald), you cannot use MLE with the zeros, but you can use GLM with zeros.

Buntin and Zaslavsky (2004) suggest that the choice and the specifics for each approach depend on the application. They provide an approach to finding a better-fitting model using a set of diagnostics from both the literature on risk adjustment and on model selection from the healthcare expenditure literature. The choice of approach appears to depend on the fraction of zeros in the data.

Finally, a one-equation alternative to two-part and selection models that can be fit with OLS yet respects the nonnegativity of the outcome variable adds a positive constant to the outcome before taking the natural logarithm. We do not recommend this approach given all the, far superior, alternatives we have described.

7.7   Statistical tests

All the usual statistical tests for single-equation models apply to the two-part model. In addition, the modified Hosmer–Lemeshow test applies to the entire two-part model. This may help identify problems with the model specification in the combined model. We can apply Pregibon's link test and Ramsey's regression equation specification error test equation by equation in these models. For the two-part model, there are no encompassing link or regression equation specification error tests, because those tests are for single-equation models. They can be extended to selection models and generalized tobit models, because they are a system of equations that can be estimated in a single MLE formulation. Pearson tests and Copas' tests can apply to all of these models.

7.8   Stata resources

The recently developed `twopm` command will not only estimate many different versions of the two-part model—allowing several options for choice of specific model—but also compute predictions and full marginal effects, accounting for retransformations, nonnormality, and heteroskedasticity (Belotti et al. 2015). Install this package by typing `ssc install twopm`. Alternatively, you can fit two-part models in two separate commands. For example, estimate the first part with either `logit` or `probit`. Commonly used commands for the second part include `regress`, `boxcox`, and `glm`—always estimated on the subsample of the data with positive values.

The Stata command for the Heckman selection model is `heckman`. A related model, with a binary equation in the second step, can be estimated with `heckprob`. Use the `tobit` command for basic tobit models with upper and lower censoring, when the censoring points are the same for all observations. Stata has two commands for generalized versions of the tobit model. Use `cnreg` when the upper- or lower-censoring points differ across observations. Use `intreg` for data that also may have interval data, in addition to point data and left- and right-censoring points.

The `treatreg` command in Stata fits a model similar to the classic selection model, but the main outcome is observed for all observations. Therefore, the selection equation can be thought of as selection into a treatment or control group. In other words, the treatment variable is a dummy endogenous variable. The model is similar to two-stage least squares, except that the errors are modeled as jointly normally distributed. Identification comes from both instruments and nonlinearities in the selection equation.

8 Count models

8.1 Introduction

In many statistical contexts, including many measures of healthcare use, the outcome or response variable of interest, y, is a nonnegative integer or count variable. Examples of count outcomes include the number of visits to the doctor, the number of nights spent in a hospital, the number of prescriptions filled, and the number of cigarettes smoked per day. Such variables have distributions that place probability mass at nonnegative integer values only. In addition, these variables are typically severely skewed, intrinsically heteroskedastic, and have variances that increase with the mean. For most count outcomes, the observations are concentrated on a few small discrete values—typically zero—and a few small positive integers. Therefore, discrete data density is an important distinguishing feature of such outcomes.

If a researcher is interested only in the prediction of the conditional mean or in the response of the conditional mean to a covariate, it may be tempting to ignore the discreteness and skewness and simply estimate the responses of interest using linear regression methods (see chapter 4). If skewness is a concern, but not discreteness, a researcher might consider the use of generalized linear models (GLMs) (see chapter 5). If discreteness and skewness are both important features of the distribution of the count outcome, then models that ignore discreteness can be quite inefficient, leading to substantial losses in statistical power. In addition, models that ignore discreteness may display considerably greater sample-to-sample variability of estimates than count models that account for the discreteness of the outcomes.

Consider a data-generating process in a regression context with a linear index, an exponential conditional mean, and a Poisson distribution. If this process leads to a distribution that is skewed and concentrated on relatively few integer values, then King (1988) demonstrated that ordinary least squares (OLS) can be grossly misspecified and produce inconsistent estimates. In addition, the OLS model on a log-transformed dependent variable (with an adjustment to account for the logarithm of zero) also produces inconsistent estimates. Figure 8.1 below illustrates such an outcome. It represents observations drawn from a Poisson distribution with a mean of 0.5. Over 90% of the values are concentrated on 0 and 1 and the distribution is distinctly skewed.

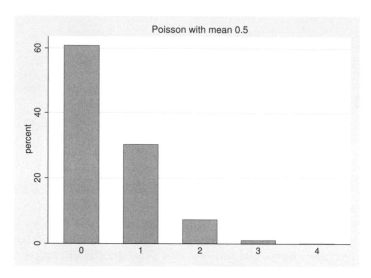

Figure 8.1. Poisson density with mean $= 0.5$

King (1988) argues that in such cases, the conditional expectation function in a count-data process cannot be linear, or even necessarily approximately linear, because predictions must be nonnegative. A regression of $\ln(y_i + c)$ on a vector of covariates **x**—where c is a small positive constant—resolves the issue of negative predictions, but King (1988) shows that results can be quite sensitive to the choice of c. In addition, Monte Carlo experiments show that the OLS estimates of such a specification are biased in both small and large samples. Furthermore, the efficiency losses are large; the Poisson MLE is 3.03 to 14.19 times more efficient than the OLS estimator of the logged, adjusted outcome.

Equally important is the consideration that in the case of discrete data, substantive interest may lie in the estimation of event probabilities. For example, researchers may wish to estimate the probability that the count equals 0, that the count is greater than 10, or that the count is greater than 2 but less than 6. There may be interest in the response of specific parts of the distribution to changes in covariates. More generally, the researcher may be interested in fitting the distribution of the event counts and examining responses of the distribution to changes in covariates, not simply in features of the conditional mean. In these situations, it is essential to formally estimate the count-data process.

Leaving aside the objective of estimating event probabilities and distributions for a moment, it is important to recognize that not all count data densities are skewed, nor is the mass concentrated on a few values in all cases—although such cases will be rare in the healthcare context. In such cases, it may well be appropriate to use methods designed for continuous outcomes. Consider the density of a random variable drawn from a Poisson distribution with a mean of five. The distribution of observations shown in figure 8.2 is relatively symmetric, so simpler models may be acceptable. Indeed, King

(1988) notes that when y_i is large for all or nearly all observations, "it would be possible to analyze this sort of data by linear least-squares techniques".

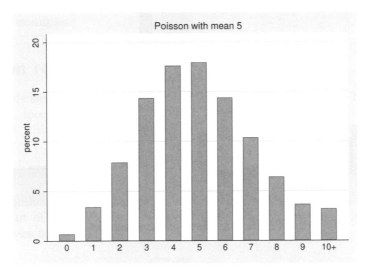

Figure 8.2. Poisson density with mean = 5

In terms of empirical regularities, it is useful to note that most of the measures of healthcare use in the 2004 Medical Expenditure Panel Survey (MEPS) dataset have probability mass concentrated on a few values and are severely skewed. We display the distributions of office visits and emergency room (ER) use in figure 8.3.

The Poisson regression model is typically fit using MLE. Given (8.1) and (8.2) and the assumption that the observations $(y_i|\mathbf{x}_i)$ are independent over i, the log-likelihood function for a sample of data can be written as

$$\ln L(\boldsymbol{\beta}) = \sum_{i=1}^{n} \{y_i \mathbf{x}_i'\boldsymbol{\beta} - \exp(\mathbf{x}_i'\boldsymbol{\beta}) - \ln(y_i!)\}$$

The first and second derivatives of the log-likelihood function, with respect to parameters $\boldsymbol{\beta}$, can be derived as

$$\frac{\partial \ln L(\boldsymbol{\beta})}{\partial \boldsymbol{\beta}} = \sum_{i=1}^{n} \{y_i - \exp(\mathbf{x}_i'\boldsymbol{\beta})\} \mathbf{x}_i$$

and

$$\frac{\partial^2 \ln L(\boldsymbol{\beta})}{\partial \boldsymbol{\beta}\partial \boldsymbol{\beta}'} = \sum_{i=1}^{n} \exp(\mathbf{x}_i'\boldsymbol{\beta}) \mathbf{x}_i \mathbf{x}_i'$$

Therefore, the Poisson MLE is the solution to k (number of parameters to be estimated) nonlinear equations corresponding to the first-order conditions for the MLE,

$$\sum_{i=1}^{n} \{y_i - \exp(\mathbf{x}_i'\boldsymbol{\beta})\} \mathbf{x}_i = 0 \tag{8.3}$$

If $\mathbf{x}_i$ includes a constant term, then the residuals $y_i - \exp(\mathbf{x}_i'\boldsymbol{\beta})$ sum to zero by (8.3).

The log-likelihood function is globally concave; hence, solving these equations by Gauss–Newton or Newton–Raphson iterative algorithms yields unique parameter estimates. By maximum likelihood theory, the estimated parameters are consistent and asymptotically normal with covariance matrix

$$\mathbf{V}\left(\widehat{\boldsymbol{\beta}}\right) = \left\{\sum_{i=1}^{n} \exp\left(\mathbf{x}_i'\boldsymbol{\beta}\right) \mathbf{x}_i \mathbf{x}_i'\right\}^{-1} \tag{8.4}$$

8.2.2 Robustness of the Poisson regression

Recall that the Poisson distribution is a member of the LEF of distributions. Therefore, the first-order conditions for the Poisson regression model MLE can be obtained from an objective function that specifies only the first moment, without specification of the distribution of the data. More precisely, it can be obtained via a GLM objective function (McCullagh and Nelder 1989) or via a pseudolikelihood (Gourieroux, Monfort, and Trognon 1984a,b) . This is clear from inspection of the first-order conditions in (8.3), because the left-hand side of (8.3) will have an expected value of zero if $E(y_i|\mathbf{x}_i) = \exp(\mathbf{x}_i'\boldsymbol{\beta})$. Therefore, parameter estimates from the Poisson regression are consistent under the relatively weak assumption that the conditional mean is correctly specified. The data-generating process need not be Poisson at all. Consequently, Poisson regression is a powerful tool for analyzing count data.

However, the standard errors of the estimates obtained by MLE are incorrect. The correct formula under the weaker assumption is

$$\mathbf{V}\left(\widehat{\boldsymbol{\beta}}\right) = \left(\sum_{i=1}^{n} \mu_i \mathbf{x}_i \mathbf{x}_i'\right)^{-1} \left\{\sum_{i=1}^{n} (y_i - \mu_i)^2 \mathbf{x}_i \mathbf{x}_i'\right\} \left(\sum_{i=1}^{n} \mu_i \mathbf{x}_i \mathbf{x}_i'\right)^{-1} \tag{8.5}$$

where $\mu_i = \exp(\mathbf{x}_i'\boldsymbol{\beta})$. This formula generally produces more conservative inference than the formula based on the MLE. Therefore, it is common practice, when estimating Poisson regressions, to implement the robust sandwich estimator of the variance (8.5) because it is appropriate under more general conditions than the maximum likelihood-based formula (8.4).

More substantively—from the point of view of an applied researcher—although we can obtain consistent estimates under weak assumptions, estimates from the Poisson regression may be grossly inefficient if the data-generating process is not Poisson. In addition, predicted probabilities and predictions of effects can be quite misleading, as we demonstrate below in the context of our data from MEPS.

8.2.3   Interpretation

The exponential mean specification of the Poisson regression model has implications for interpreting the parameters of the model. As with all exponential mean models, the coefficients themselves have a natural interpretation as semielasticities with respect to the variables (or elasticity if the variable itself is measured in logarithms). More precisely, because $E(y_i|\mathbf{x}_i) = \exp(\mathbf{x}_i'\boldsymbol{\beta})$,

$$\frac{\partial \ln\left\{E\left(y_i|\mathbf{x}_i\right)\right\}}{\partial x_j} = \beta_j$$

where the scalar, x_j, denotes the jth regressor. To demonstrate the interpretation of results from the Stata output, we display the Stata log from a Poisson regression of office-based visits (`use_off`) on a simple specification of covariates that includes one continuous variable (`age`) and one binary indicator (`female`). Note that an increase in age by 1 year leads to 2.5% more visits [$100\{\exp(0.0242)-1\}$]. In addition, women have about 50% (derived from the point estimate on `1.female` using $100\{\exp(0.406) - 1\}$) more visits than men.

depends on the range and distribution of the covariate and on the estimate of the associated parameter.

8.2.4   Is Poisson too restrictive?

Recall that the Poisson distribution is parameterized in terms of a single scalar parameter (μ), so all moments of y are functions of μ. In fact, both the mean and variance of a Poisson random variable are μ. In spite of this seemingly draconian restriction, we have shown that parameter estimates from the Poisson regression are consistent, even when the data-generating process is not Poisson—that is, this equality property does not hold.

As we have seen in the MEPS dataset, empirical distributions of healthcare use are overdispersed relative to the Poisson—that is, the variance exceeds the mean. Overdispersion leads to deflated standard errors and inflated t statistics in the usual maximum likelihood output. Ignoring the overdispersion will lead to a false sense of precision, and the greater the discrepancy between the variance and the mean, the more the risk grows. However, this issue can be remedied with robust standard errors estimated using "sandwich" estimators of the variance–covariance matrix of parameters. Note that obtaining correct standard-errors does not render the estimates efficient. Poisson MLE is still inefficient and can be grossly so.

The specification below estimates a sandwich variance–covariance matrix of parameter estimates and reports robust standard errors in the case of the MEPS data for the count of office visits. Comparing the output below to the estimates obtained without vce(robust) shows that the standard errors of the coefficients are approximately four times larger using the vce(robust) option. This example demonstrates the inefficiency of the Poisson estimator for such counts and highlights the importance of using robust standard errors for inference if Poisson is the desired estimator.

```
. *** Poisson coefficients with robust standard errors
. poisson use_off age i.female, vce(robust)

Iteration 0:   log pseudolikelihood =  -117955.2
Iteration 1:   log pseudolikelihood = -117955.19
Iteration 2:   log pseudolikelihood = -117955.19

Poisson regression                              Number of obs    =      19,386
                                                Wald chi2(2)     =     1874.25
                                                Prob > chi2      =      0.0000
Log pseudolikelihood = -117955.19               Pseudo R2        =      0.1023
```

use_off	Coef.	Robust Std. Err.	z	P>|z|	[95% Conf. Interval]	
age	.0242174	.0006212	38.99	0.000	.0229999	.0254349
female						
Female	.4055818	.0280805	14.44	0.000	.3505451	.4606185
_cons	.3201797	.0384744	8.32	0.000	.2447714	.3955881

Finally, even though parameter estimates are consistent, estimates of marginal and incremental effects and event probabilities can be inconsistent. For example, the Poisson density often underpredicts event probabilities in both tails of the distribution. We first reestimate a Poisson regression for office-based visits and calculate the observed and predicted probabilities using the Stata code shown below. The predicted density is calculated for each value of the count variable (up to a maximum value based on the empirical frequency for each outcome) and for each observation (that is, for different values of covariates). Then, the predicted frequencies are averaged to obtain a single measure of the average predicted density for each count value.

```
. *** Poisson regression observed and predicted densities
. poisson use_off age i.female

Iteration 0:    log likelihood =  -117955.2
Iteration 1:    log likelihood = -117955.19
Iteration 2:    log likelihood = -117955.19

Poisson regression                              Number of obs   =      19,386
                                                LR chi2(2)      =    26894.76
                                                Prob > chi2     =      0.0000
Log likelihood = -117955.19                     Pseudo R2       =      0.1023

-------------------------------------------------------------------------------
     use_off |      Coef.   Std. Err.      z    P>|z|     [95% Conf. Interval]
-------------+-----------------------------------------------------------------
         age |   .0242174   .0001637   147.97   0.000     .0238966    .0245381
             |
      female |
      Female |   .4055818   .0062865    64.52   0.000     .3932604    .4179032
       _cons |   .3201797   .0099091    32.31   0.000     .3007583    .3396012
-------------------------------------------------------------------------------

. forvalues j=0/20 {
  2.          generate byte y_`j' = `e(depvar)' == `j'
  3.          predict pr_`j', pr(`j')
  4. }
```

We also calculate observed and predicted probabilities for a Poisson model of the count of ER visits in an analogous fashion. The distributions are displayed in figure 8.4. The light (open) bars depict the empirical density, that is, the frequency of observations in each count cell. The dark bars depict the predicted frequencies. The figure highlights the extent to which event probabilities of tail events are underpredicted, especially the zeros; consequently, events in the center of the distribution are overpredicted for the model of office-based visits. The Poisson assumption appears a better fit for ER use.

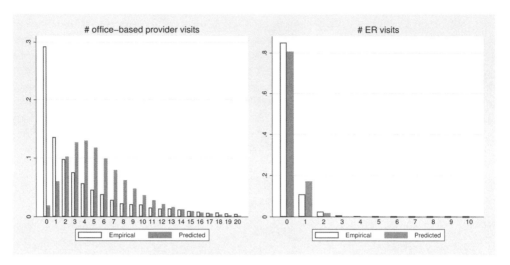

Figure 8.4. Empirical and Poisson-predicted frequencies

8.3   Negative binomial models

The negative binomial regression model is arguably the most popular model for count data that accommodates overdispersion. It is often justified as a logical extension of the Poisson regression in which overdispersion (relative to the Poisson) is caused by unobserved heterogeneity. Consider an unobserved random term in the conditional mean specification in the Poisson regression—that is, $\mu_i = \exp(\mathbf{x}_i'\boldsymbol{\beta} + \xi_i)$. In the context of models of health and healthcare use, it is not hard to justify ξ_i via the existence of unobserved differences in health status or differences in tastes. The former is especially appealing in the absence of rich specifications of health status—or the observations that many chronic conditions that affect utilization are relatively rare, and their severity is rarely measured. Integration of ξ_i out of the distribution leads to the negative binomial distribution. This, and other derivations of the negative binomial distribution, is in Cameron and Trivedi (2013).

The negative binomial density for a count outcome $y = 0, 1, 2, 3, \dots$ is

$$f(y_i; \mu, \alpha) = \frac{\Gamma(\alpha^{-1} + y_i)}{\Gamma(\alpha^{-1})\Gamma(y_i + 1)} \left(\frac{\alpha^{-1}}{\alpha^{-1} + \mu} \right)^{\alpha^{-1}} \left(\frac{\mu}{\alpha^{-1} + \mu} \right)^{y_i}, \quad \alpha > 0 \qquad (8.6)$$

where $\Gamma(\cdot)$ denotes the gamma function that simplifies to a factorial for an integer argument, α is an additional parameter, and μ has the same interpretation as in the Poisson model.

The first two moments of the negative binomial distribution are

$$
\begin{aligned}
E(y_i; \mu, \alpha) &= \mu \\
\mathbf{V}(y_i; \mu, \alpha) &= \mu + \alpha\mu^2
\end{aligned}
\qquad (8.7)
$$

An appealing property of this parameterization is that the conditional mean of the negative binomial regression is exactly the same as that in the Poisson regression. However, the variance exceeds the mean. Thus the negative binomial distribution introduces a greater proportion of zeros and a thicker right tail. Figure 8.5 displays histograms of Poisson and negative binomial densities with means of two. A researcher can visually observe that the negative binomial density is overdispersed relative to the Poisson and has considerably larger fractions of zeros and "large" (greater than 10) values.

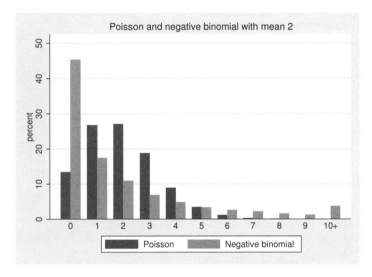

Figure 8.5. Negative binomial density

Two standard variants of the negative binomial are used in regression applications. Both variants specify conditional means using $\mu_i = \exp(\mathbf{x}_i'\boldsymbol{\beta})$. The most common variant specifies the conditional variance of y_i as $\mu_i + \alpha\mu_i^2$ [from (8.7)], which is quadratic in the mean. The other variant of the negative binomial model specifies the variance $\mu_i + \delta\mu_i$, which is linear in the mean. This specification is derived by replacing α with δ/μ_i throughout (8.6). In Cameron and Trivedi (2013) and elsewhere, this formulation is called the negative binomial-1 (NB1) model, while the formulation with the quadratic variance function is called the negative binomial-2 (NB2) model.

The negative binomial distribution is not a member of the LEF, so it is sensitive to misspecification. Unlike with the Poisson distribution, one must be sure that the data-generating process is a negative binomial to ensure that the parameter estimates are consistent. However, Cameron and Trivedi (2013) show that a negative binomial regression model with a fixed value of α (or δ) has a distribution in the LEF and hence is robust to misspecification of higher moments. Because of this property, negative binomial regression estimates are quite reliable in practice.

8.3.1   Examples of negative binomial models

The examples below show the results of NB1 and NB2 models fit for the count of the number of office-based visits. The NB2 regression is the default specification of the nbreg command in Stata (or the dispersion(mean) option), so we fit that model first. Parameter estimates from negative binomial regressions have a semielasticity interpretation. We see that the effect of an additional year in age is associated with a 2.8% increase in office visits. Women have 52% more visits than men.

```
. *** NB2 regression coefficients
. use http://www.stata-press.com/data/heus/heus_mepssample, clear
(Sample of MEPS 2004 data)

. nbreg use_off age i.female, vce(robust)

Fitting Poisson model:

Iteration 0:   log pseudolikelihood =  -117955.2
Iteration 1:   log pseudolikelihood = -117955.19
Iteration 2:   log pseudolikelihood = -117955.19

Fitting constant-only model:

Iteration 0:   log pseudolikelihood = -55053.853
Iteration 1:   log pseudolikelihood = -52509.158
Iteration 2:   log pseudolikelihood = -52504.625
Iteration 3:   log pseudolikelihood = -52504.625

Fitting full model:

Iteration 0:   log pseudolikelihood = -51498.113
Iteration 1:   log pseudolikelihood = -51340.252
Iteration 2:   log pseudolikelihood = -51332.107
Iteration 3:   log pseudolikelihood = -51332.097
Iteration 4:   log pseudolikelihood = -51332.097
```

Negative binomial regression				Number of obs	=	19,386
				Wald chi2(2)	=	1509.90
Dispersion	= mean			Prob > chi2	=	0.0000
Log pseudolikelihood = -51332.097				Pseudo R2	=	0.0223

use_off	Coef.	Robust Std. Err.	z	P>|z|	[95% Conf. Interval]	
age	.027692	.0007396	37.44	0.000	.0262423	.0291417
female						
Female	.5192054	.0300945	17.25	0.000	.4602212	.5781896
_cons	.0802363	.0476959	1.68	0.093	-.0132459	.1737186
/lnalpha	.652287	.0154658			.6219746	.6825994
alpha	1.919927	.0296932			1.862602	1.979015

As always, it is useful to compute effect sizes on the natural scale, so we use margins to calculate sample average marginal and incremental effects. The results show that individuals who are a year older have 0.16 more visits on average. Women have 2.9 more visits than men.

```
. *** NB2 regression marginal effects
. margins, dydx(age female)
Average marginal effects                          Number of obs    =      19,386
Model VCE       : Robust

Expression      : Predicted number of events, predict()
dy/dx w.r.t.    : age 1.female
```

	dy/dx	Delta-method Std. Err.	z	P>|z|	[95% Conf. Interval]	
age	.1645081	.0053657	30.66	0.000	.1539915	.1750247
female Female	2.925267	.1633858	17.90	0.000	2.605036	3.245497

```
Note: dy/dx for factor levels is the discrete change from the base level.
```

Next, we estimate the NB1 regression, which requires the `dispersion(constant)` option. Parameter estimates are below. Note that Stata reports a parameter, `alpha`, which is the value of exponentiated `/lnalpha` in the case of the NB2 regression. In the case of the NB1 regression, Stata reports a parameter, `delta`, which is the value of the exponentiated `/lndelta`. The coefficient estimates on age are roughly similar across NB2 and NB1 specifications, while the coefficient on female is smaller when fit using the NB1 model.

```
. *** NB2 regression coefficients
. nbreg use_off age i.female, dispersion(constant) vce(robust)
Fitting Poisson model:

Iteration 0:    log pseudolikelihood =  -117955.2
Iteration 1:    log pseudolikelihood = -117955.19
Iteration 2:    log pseudolikelihood = -117955.19

Fitting constant-only model:

Iteration 0:    log pseudolikelihood = -79199.366
Iteration 1:    log pseudolikelihood = -54709.926
Iteration 2:    log pseudolikelihood = -52597.055
Iteration 3:    log pseudolikelihood = -52504.675
Iteration 4:    log pseudolikelihood = -52504.625
Iteration 5:    log pseudolikelihood = -52504.625

Fitting full model:

Iteration 0:    log pseudolikelihood = -52504.625
Iteration 1:    log pseudolikelihood = -51187.223
Iteration 2:    log pseudolikelihood = -50771.465
Iteration 3:    log pseudolikelihood =  -50768.74
Iteration 4:    log pseudolikelihood = -50768.739
```

```
Negative binomial regression                    Number of obs    =    19,386
                                                Wald chi2(2)     =   4391.72
Dispersion              = constant              Prob > chi2      =    0.0000
Log pseudolikelihood = -50768.739              Pseudo R2        =    0.0331
```

use_off	Coef.	Robust Std. Err.	z	P>|z|	[95% Conf. Interval]	
age	.0230037	.0004086	56.30	0.000	.0222029	.0238045
female						
Female	.4153586	.0154408	26.90	0.000	.3850952	.4456219
_cons	.3782391	.0267831	14.12	0.000	.3257453	.4307329
/lndelta	2.372466	.023307			2.326785	2.418146
delta	10.7238	.2499398			10.24495	11.22503

Estimates of the sample average partial effects reveal that both the marginal effect of age (0.13) and the incremental effect of female (2.32) are smaller when fit using the NB1 model.

```
. *** NB2 regression coefficients
. margins, dydx(age female)
Average marginal effects                        Number of obs    =    19,386
Model VCE      : Robust

Expression    : Predicted number of events, predict()
dy/dx w.r.t.  : age 1.female
```

	dy/dx	Delta-method Std. Err.	z	P>|z|	[95% Conf. Interval]	
age	.1334761	.0028539	46.77	0.000	.1278827	.1390696
female						
Female	2.319704	.0868393	26.71	0.000	2.149502	2.489906

Note: dy/dx for factor levels is the discrete change from the base level.

The NB1 and NB2 models are not nested models—so in principle, a researcher should use tests to discriminate among nonnested models such as Vuong's (1989) test or model-selection criteria such as the Akaike information criterion (AIC) or the Bayesian information criterion (BIC) (see chapter 2). But because the NB1 and NB2 models have the same number of parameters, most nonnested tests and criteria simplify to a comparison of maximized log likelihoods. The value of the log likelihoods suggests that NB1 fits better than NB2 for this particular dataset and model specification.

As we did with the Poisson regressions, we calculate empirical frequencies of each count value and the associated predicted frequencies from NB2 regressions for office-based visits and ER use. The histograms of actual and predicted count frequencies are shown in figure 8.6. The left panel demonstrates the dramatic improvement in fit of the NB2 regression relative to Poisson for office-based provider visits (compare with the left panel of figure 8.4). The improvement in fit for ER visits is not as dramatic but still noticeable.

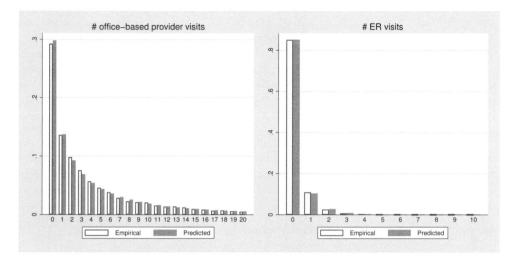

Figure 8.6. Empirical and NB2 predicted frequencies

One way to improve the fit of the negative binomial model even further involves a parameterization of α in terms of a linear combination of a set of covariates. However, although this extended model is more flexible in principle, the parameters of such models can be difficult to identify in finite samples. Instead, in the following sections, we describe two types of models that add flexibility to the basic Poisson or negative binomial specifications, have considerable intuitive appeal, and often fit the counts of healthcare use quite well.

8.4   Hurdle and zero-inflated count models

As mentioned above, many counts of health and healthcare use have more zeros than predicted by Poisson or negative binomial models. Thus the first extensions we consider to the Poisson and negative binomial regressions are their hurdle and zero-inflated extensions. Each of these models adds flexibility by relaxing the assumption that the zeros and the positives come from the same data-generating process. Each naturally generates excess zeros and a thicker right tail relative to the parent distributions, but they are also capable of generating fewer zeros and thinner tails.

8.4.1   Hurdle count models

The hurdle count model can have the same conceptual justification as is often used to justify the two-part model—that it reflects a two-part decision-making process (see also chapter 7). One motivation is based on a principal-agent mechanism. First, the principal decides whether to use the medical care or not. Then—conditional on making the decision to use care—the agent, on behalf of the principal, makes a second decision about how much care to consume. More specifically, the patient initiates the first visit to a doctor, but the doctor and patient jointly determine the second and subsequent visits (Pohlmeier and Ulrich 1995). Alternatively, the two-step process could be thought of as driven by transaction costs of entry into the market, which do not exist once the individual is engaged in the receipt of healthcare services. A richer formulation of the principal-agent mechanism models the fact that bouts of illness arise during the course of the year. Some factors may have a differential effect on whether these episodes of illness become episodes of treatment—for example, the opportunity to visit one's family physician rather than having to go to an ER (Keeler and Rolph 1988).

However, such justifications are not required for the hurdle count model to be an appealing extension to the standard Poisson and negative binomial model. Instead, it is enough to acknowledge that there may be substantial heterogeneity at the threshold of the count variable between use and nonuse.

In the hurdle, or two-part model, the zeros are determined by one density, $f_1(\cdot)$, so that $\Pr(y_i = 0) = f_1(0)$—while the positive counts are from another density, $f_2(\cdot)$. To be more precise, the positive counts are drawn from the truncated density, $f_2(y_i|y_i > 0) = f_2(y_i)/\{1 - f_2(0)\}$. Section 8.5 provides more details on truncated counts. The overall data-generating mechanism is

$$g(y_i|\mathbf{x}_i; \boldsymbol{\theta_1}, \boldsymbol{\theta_2}) = \begin{cases} f_1(0|\mathbf{x}_i; \boldsymbol{\theta_1}) & \text{if } y_i = 0 \\ \frac{1-f_1(0|\mathbf{x}_i; \boldsymbol{\theta_1})}{1-f_2(0|\mathbf{x}_i; \boldsymbol{\theta_2})} f_2(y_i|\mathbf{x}_i; \boldsymbol{\theta_2}) & \text{if } y_i = 1, 2, 3, \ldots \end{cases}$$

In practice, $f_1(\cdot)$ is usually specified as a logit or probit, although any binary choice model will do. The distribution, $f_2(\cdot)$, is usually a Poisson or negative binomial. When $f_2(\cdot)$ is a negative binomial, the hurdle count model takes overdispersion into account in two ways: by allowing a separate process for the zeros, and by allowing the positive counts to be overdispersed via the negative binomial parameter, α.

To demonstrate, we return to the example of any office visits with the MEPS data. There are two parts to hurdle Poisson model estimation. In the first step, we fit a logit model for the probability that the number of office-based visits is greater than zero. Age and female both increase the probability of having at least one office-based visit significantly.

```
. *** Hurdle Poisson model hurdle part estimates
. generate any_off = use_off>0

. logit any_off age i.female, vce(robust)
Iteration 0:   log pseudolikelihood = -11718.705
Iteration 1:   log pseudolikelihood = -10464.079
Iteration 2:   log pseudolikelihood = -10420.727
Iteration 3:   log pseudolikelihood = -10420.605
Iteration 4:   log pseudolikelihood = -10420.605

Logistic regression                             Number of obs    =      19,386
                                                Wald chi2(2)     =     2092.96
                                                Prob > chi2      =      0.0000
Log pseudolikelihood = -10420.605               Pseudo R2        =      0.1108
```

any_off	Coef.	Robust Std. Err.	z	P>|z|	[95% Conf. Interval]	
age	.0447412	.0010934	40.92	0.000	.0425982	.0468842
female						
Female	.9086022	.034277	26.51	0.000	.8414206	.9757838
_cons	-1.482785	.0519409	-28.55	0.000	-1.584587	-1.380983

The estimates of marginal effects show that women are almost 17 percentage points more likely to have at least one office-based visit than men. An extra year in age leads to a 0.8 percentage point increase in the probability of at least one office-based visit.

```
. *** Hurdle Poisson model hurdle part marginal effects
. margins, dydx(*)
Average marginal effects                        Number of obs    =      19,386
Model VCE    : Robust

Expression   : Pr(any_off), predict()
dy/dx w.r.t. : age 1.female
```

	dy/dx	Delta-method Std. Err.	z	P>|z|	[95% Conf. Interval]	
age	.0080721	.0001701	47.47	0.000	.0077388	.0084054
female						
Female	.1691545	.0061898	27.33	0.000	.1570228	.1812862

Note: dy/dx for factor levels is the discrete change from the base level.

In the second step, we fit a truncated Poisson model using the Stata command `tpoisson`. For estimation, it is important to condition on the sample with positive values for the outcome; that is, drop observations with zero counts. Notice the use of the `if use_off>0` qualifier in the `tpoisson` command shown below:

```
. *** Hurdle Poisson model estimates
. tpoisson use_off age i.female if use_off>0, ll(0) vce(robust)

Iteration 0:   log pseudolikelihood =  -88051.84
Iteration 1:   log pseudolikelihood = -88051.696
Iteration 2:   log pseudolikelihood = -88051.696

Truncated Poisson regression                    Number of obs   =      13,713
Limits:         lower =        0               Wald chi2(2)    =      635.31
                upper =     +inf               Prob > chi2     =      0.0000
Log pseudolikelihood = -88051.696              Pseudo R2       =      0.0476
```

use_off	Coef.	Robust Std. Err.	z	P>|z|	[95% Conf. Interval]	
age	.0147923	.0006073	24.36	0.000	.0136019	.0159826
female						
Female	.2108255	.0268828	7.84	0.000	.1581361	.2635149
_cons	1.215349	.0397453	30.58	0.000	1.137449	1.293248

A variety of marginal effects can be calculated using `margins`. We can get estimates of marginal effects on the conditional (on the outcome being greater than zero) mean for the entire sample, not just for observations with the outcome greater than zero. Conditional on having at least one office-based visit—women average 1.59 visits more than men. On average, a person who is one year older is expected to have 0.11 more visits.

```
. *** Hurdle Poisson model marginal effects
. margins, dydx(*) predict(cm) noesample

Average marginal effects                        Number of obs   =      19,386
Model VCE    : Robust

Expression   : Conditional mean of n > ll(0), predict(cm)
dy/dx w.r.t. : age 1.female
```

	dy/dx	Delta-method Std. Err.	z	P>|z|	[95% Conf. Interval]	
age	.1133792	.0046826	24.21	0.000	.1042014	.1225569
female						
Female	1.591346	.1963381	8.11	0.000	1.20653	1.976161

Note: dy/dx for factor levels is the discrete change from the base level.

Predictions and marginal effects from the hurdle Poisson model, as a whole, require putting the two parts together. Without the convenience of a command such as `twopm`, the estimates can be combined using `suest` and the `expression()` option in `margins` to obtain overall marginal effects. We first refit the logit and truncated Poisson regression models without adjustments to the maximum-likelihood standard errors, because those adjustments are done within `suest`. `suest` produces a typical Stata regression table with coefficients and standard errors of both equations.

```
. *** Hurdle Poisson model suest
. quietly logit any_off age i.female

. estimates store h1

. quietly tpoisson use_off age i.female if use_off>0, ll(0)

. estimates store h2

. suest h1 h2
Simultaneous results for h1, h2

                                             Number of obs       =      19,386
```

	Coef.	Robust Std. Err.	z	P>|z|	[95% Conf. Interval]	
h1_any_off						
age	.0447412	.0010934	40.92	0.000	.0425982	.0468842
female						
Female	.9086022	.034277	26.51	0.000	.8414206	.9757838
_cons	-1.482785	.0519409	-28.55	0.000	-1.584587	-1.380983
h2_use_off						
age	.0147923	.0006073	24.36	0.000	.0136019	.0159826
female						
Female	.2108255	.0268825	7.84	0.000	.1581367	.2635143
_cons	1.215349	.0397448	30.58	0.000	1.13745	1.293247

Next, we code the formula for the conditional mean of the outcome for the hurdle Poisson model and pass that to the expression() option of the margins command. The code and results are shown below. They show that women have 2.36 more office-based visits than men. From the previous results, we can conclude that this is because they are more likely to have an office-based visit and because, among those who visit, they have more visits. An extra year in age increases the number of visits by 0.14, again because of increases along extensive and intensive margins.

```
. *** Hurdle Poisson model marginal effects
. local logit "invlogit(predict(eq(h1_any_off))) "

. local ey "exp(predict(eq(h2_use_off))) "

. local pygt0 "poissontail(exp(predict(eq(h2_use_off))),1)"
```

```
. margins, dydx(*) expression("`logit´*`ey´/`pygt0´")
Warning: cannot perform check for estimable functions.
Average marginal effects                          Number of obs    =      19,386
Model VCE     : Robust
Expression    : invlogit(predict(eq(h1_any_off)))
                *exp(predict(eq(h2_use_off)))
                /poissontail(exp(predict(eq(h2_use_off))),1)
dy/dx w.r.t. : age 1.female
```

	dy/dx	Delta-method Std. Err.	z	P>|z|	[95% Conf. Interval]	
age	.1415167	.0038429	36.83	0.000	.1339847	.1490487
female						
Female	2.359174	.1488385	15.85	0.000	2.067456	2.650892

```
Note: dy/dx for factor levels is the discrete change from the base level.
```

We also fit a truncated negative binomial regression model, combine estimates from the two parts, and estimate marginal effects using the same steps as for the hurdle Poisson model. The sample average of the incremental effect of being female is 2.54: women average 2.54 more office-based visits than men. This estimate is somewhat larger than the one obtained from the hurdle Poisson. An extra year of age is estimated to increase the number of visits by 0.15, which is quite similar to that obtained from the hurdle Poisson.

```
. *** Hurdle NB2 model marginal effects
. quietly logit any_off age i.female
. estimates store h1
. quietly tnbreg use_off age i.female if use_off>0, ll(0)
. estimates store h2
. quietly suest h1 h2
.
. local logit "invlogit(predict(eq(h1_any_off))) "
. local ey "exp(predict(eq(h2_use_off))) "
. local pygt0 "(nbinomialtail(exp(-predict(eq(/h2:lnalpha))),1,"
>         "1/(1+exp(predict(eq(h2_use_off)))/exp(-predict(eq(/h2:lnalpha)))))))"
```

```
. margins, dydx(*) expression("`logit´*`ey´/`pygt0´")
Warning: cannot perform check for estimable functions.
Average marginal effects                      Number of obs    =     19,386
Model VCE    : Robust

Expression   : invlogit(predict(eq(h1_any_off)))
               *exp(predict(eq(h2_use_off)))
               /(nbinomialtail(exp(-predict(eq(/h2:lnalpha))),1,
               1/(1+exp(predict(eq(h2_use_off))))/exp(-predict(eq(/h2:lnalpha)
               > )))))
dy/dx w.r.t. : age 1.female
```

	dy/dx	Delta-method Std. Err.	z	P>|z|	[95% Conf. Interval]	
age	.1500508	.0044807	33.49	0.000	.1412688	.1588328
female						
Female	2.54379	.1561728	16.29	0.000	2.237697	2.849883

```
Note: dy/dx for factor levels is the discrete change from the base level.
```

Note that the sample average incremental and marginal effects obtained from the hurdle specification are quite close to those obtained using the standard negative binomial regression. However, this does not mean that the partial effects would be similar throughout the distribution of the covariates.

8.4.2 Zero-inflated models

The zero-inflated model developed by Lambert (1992) and Mullahy (1997) also has considerable intuitive appeal. The intuition is that there are two types of agents in the population: potential users and nonusers. While positive counts arise only from the decisions of users, zeros can arise because users choose not to consume in a particular period or because of the behavior of nonusers.

In the context of healthcare use, consider an example in which the outcome of interest is the number of visits to an acupuncturist. It might be reasonable to believe that the population consists of two types of individuals: those who would never seek such care and those who would. However, there would be individuals among the second group who did not visit an acupuncturist in the survey recall period. A person observed to have zero visits during the observation period might either be someone who would never visit an acupuncturist or be someone who would—but happened not to during the observation period. Thus a zero-inflated model would be a powerful way to model the additional heterogeneity relative to a standard model. Note that as with the hurdle count model, the use of the zero-inflated model need not be justified using the intuition of two types of individuals in the population. It may simply be used to provide additional modeling flexibility.

In the zero-inflated class of models, a count density, $f_2(\cdot)$, produces realizations from the entire range of nonnegative integers—that is, $y_i = 0, 1, 2, \ldots$. In addition, another

```
. *** Zero-inflated NB2 model margins
. margins, dydx(age female)
```

Average marginal effects   Number of obs   =   19,386
Model VCE   : Robust

Expression   : Predicted number of events, predict()
dy/dx w.r.t. : age 1.female

	dy/dx	Delta-method Std. Err.	z	P>|z|	[95% Conf. Interval]	
age	.1509562	.0045735	33.01	0.000	.1419922	.1599201
female Female	2.488619	.1594008	15.61	0.000	2.176199	2.801039

```
Note: dy/dx for factor levels is the discrete change from the base level.
```

Typically, the larger the discrepancy between the number of zeros in the data and the number predicted by the standard count model (Poisson or negative binomial), the greater the gains will be from the additional modeling complexities of either the hurdle or the zero-inflated model. Gains will be most obvious along the dimension of predicted counts, but a researcher will typically also obtain better estimates of other event probabilities and partial effects. Further, unlike the choice between Poisson and negative binomial, where the choice of distribution has no impact on expected outcome in terms of the conditional mean, the move to a zero-inflated Poisson or negative binomial regression model—or a hurdle count model—can change the conditional mean response to the covariates.

8.5 Truncation and censoring

In this section, we briefly discuss situations in which the count outcome is either truncated (missing) or censored (recoded). Truncation occurs naturally in the second part of the hurdle count model but also in situations where nonusers of care are unobserved. Censoring occurs when counts are top coded.

8.5.1 Truncation

In some studies, sampled individuals must have been engaged in the activity of interest to be included in the samples. When this is the case, the count data are truncated, because they are observed only over part of the range of the response variable. Examples of truncated counts include outpatient visits and the number of prenatal visits among a population of women who all have at least one visit. In all of these cases, we do not observe zero counts, so the data are said to be zero-truncated—or, more generally, left-truncated. Right-truncation is less common but can arise in situations where the analyst has a dataset in which observations with large values of the count variable have been removed from the dataset during prior analysis. Typically, researchers know the rule by which these observations were removed, but it is not feasible to recover those observations from the original data source. Truncation leads to inconsistent parameter estimates, unless the likelihood function is suitably modified. This is the case even when the true data-generating mechanism is as simple as a Poisson density (Gurmu 1997).

Consider the case of zero-truncation. Let $f(y_i|\mathbf{x}_i; \boldsymbol{\theta})$ denote the density function, and let $F(y_i|\mathbf{x}_i; \boldsymbol{\theta})$ denote the cumulative distribution function of the discrete random variable, where $\boldsymbol{\theta}$ is a parameter vector. If realizations of $y_i < 1$ are omitted, then the ensuing zero-truncated density is

$$f(y_i|\mathbf{x}_i; \boldsymbol{\theta}, y_i \geq 1) = \frac{f(y_i|\mathbf{x}_i; \boldsymbol{\theta})}{1 - F(0|\mathbf{x}_i; \boldsymbol{\theta})}, \quad y_i = 0, 1, 2, \ldots$$

For the zero-truncated Poisson, this simplifies to

$$f(y_i|\mathbf{x}_i; \mu_i, y_i \geq 1) = \frac{e^{-\mu_i} \mu_i^{y_i}}{y_i!(1 - e^{-\mu_i})}$$

where $\mu_i = \exp(\mathbf{x}_i\boldsymbol{\beta})$. Maximum likelihood estimates of zero-truncated models are implemented in Stata for the Poisson and negative binomial densities with the `tpoisson` and `tnbreg` commands, respectively.

8.5.2   Censoring

Censored counts most commonly arise from the aggregation of counts greater than some value. This is often done in survey designs involving healthcare use for confidentiality reasons, where a measure of use is top-coded at a value smaller than the true maximum in the data to reduce the risk of compromising confidentiality of the survey respondent. Censoring, like truncation, leads to inconsistent parameter estimates if the uncensored likelihood is mistakenly used (Gurmu 1997). Consider the case in which the number of events greater than some known value, c, might be aggregated into a single category. In this case, some values of y are incompletely observed; the precise value is unknown, but it is known to equal or exceed c. The observed data have density

$$g(y_i|\mathbf{x}_i; \boldsymbol{\theta}) = \begin{cases} f(y_i|\mathbf{x}_i; \boldsymbol{\theta}) & \text{if } y_i < c \\ 1 - F(c - 1|\mathbf{x}_i; \boldsymbol{\theta}) & \text{if } y_i \geq c \end{cases}$$

Simplification to the Poisson and negative binomial densities can be derived using their respective densities.

8.6   Model comparisons

As is apparent from the number of models and their variants described above, a number of modeling choices are necessary when modeling a count outcome. Although the Poisson regression has the advantage of being robust to some types of misspecification, the estimates from a Poisson regression may not be desirable, and predictions of events and partial effects of policy interest may well be inconsistent. The richer class of models

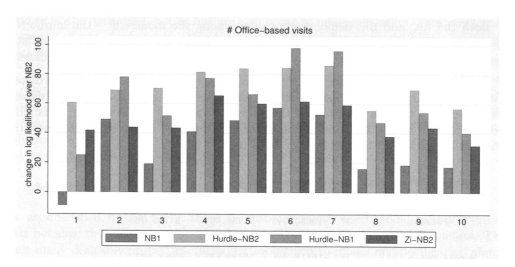

Figure 8.7. Cross-validation log likelihood for office-based visits

The evidence for ER visits, shown in figure 8.8, is quite different. There is virtually no discrimination between NB1 and NB2 models or among their extensions.

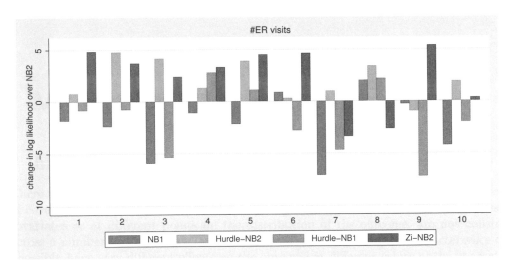

Figure 8.8. Cross-validation log likelihood for ER visits

8.7   Conclusion

We have described a number of models for count data in this chapter. They are useful as models for many measures of healthcare use. In one of the two empirical examples in this chapter, there is considerable gain in fit from going beyond the standard Poisson and negative binomial models to hurdle and zero-inflated extensions of those models. Nevertheless, even the hurdle and zero-inflated extensions may sometimes be insufficient to provide an adequate fit for some outcomes. We will return to the development of other, possibly more flexible extensions in chapter 9.

8.8   Stata resources

Stata has several commands to estimate count models as well as zero-inflated count models. To estimate the Poisson regression in Stata, use the `poisson` command. Use `nbreg` for negative binomial regressions, `zip` for the zero-inflated Poisson regression, and `zinb` for its negative binomial counterpart. Hurdle count models do not have single commands in Stata. They can be estimated by using `logit` or `probit` for the hurdle part of the model and `tpoisson` or `tnbreg` for the conditional (on a positive integer) part of the model.

To compare the AIC and BIC test statistics for the count models described above, use `estimates stats *` or `estat ic`.

There are often good reasons to believe that effects are heterogeneous along dimensions that cannot easily be characterized by parametric functional forms or by interactions of covariates as they are typically specified. For example, effects may be heterogeneous along the values of the outcome itself, by complex configurations of observed characteristics, or on unobserved characteristics. These types of effect heterogeneity are not easy to account for using the models that we have described so far. Ignoring heterogeneity may be a lost opportunity for greater understanding in some cases, while it may lead to misleading conclusions in others.

As we saw in chapter 2, accounting for such heterogeneity is exceedingly important if the researcher is interested in estimating effects at specific values of covariates or of the outcome. We also saw that allowing for appropriate nonlinearity might be important even if the object of interest is an average of effects across the sample, for example, the average treatment effect. Therefore, in this chapter, we describe four methods that allow the researcher to explore heterogeneity of effects in more general ways: First, we describe quantile regression, which is an appealing technique to explore heterogeneity along values of the outcome. Next, we describe finite mixture models, which allow for heterogeneity along the outcome distribution by complex configurations of either observed or unobserved characteristics. These models identify a finite (typically small) number of classes of observations with associated covariate effects that vary across classes. Third, we describe some uses of nonparametric regression now available in Stata. Nonparametric regression techniques make few assumptions about the functional form of the relationship between the outcome and the covariates. Finally, we briefly describe a conditional density estimator that explicitly allows the relationship between covariates and outcome to differ across the distribution of the outcome.

9.2   Quantile regression

So far, the models we have described relate the conditional mean of the outcome, $E(y|\mathbf{x})$, to a set of covariates through one (or two) linear indexes of parameters and covariates; that is, $E(y|\mathbf{x}) = g(\mathbf{x}_i'\boldsymbol{\beta})$. In quantile regression, the conditional expectation of the outcome is not modeled. Instead, the conditional τth quantile is modeled using a linear index of covariates; that is, $Q_\tau(y|\mathbf{x}) = \mathbf{x}_i'\boldsymbol{\beta}$. When $\tau = 0.5$, the quantile regression is also known as a median regression. As we will see below, quantile regressions allow effects of covariates to vary across conditional quantiles; that is, the effect of a covariate on the τ_1th conditional quantile may be different from the effect of the same covariate on the τ_2th quantile. Thus quantile regressions provide a way for researchers to understand how the effect of a covariate might differ across the distribution of the outcome.

The median regression, or the 0.5th quantile regression, is the simplest point of departure from the linear model fit by OLS. Specifically, consider the linear regression specification for the continuous dependent variable, y_i, for individuals i, regressed on a vector of covariates, $\mathbf{x}_i$—with a vector of parameters to be estimated, $\boldsymbol{\beta}$, and an independent and identically distributed error term, u_i:

$$y_i = \mathbf{x}_i'\boldsymbol{\beta} + u_i$$

In ordinary least squares (OLS), the parameters of this linear model are computed as the solution to minimizing the sum of squared residuals. In an analogous way, the median quantile regression computes parameters of the same linear regression specification by minimizing the sum of absolute residuals. The sum of squared residuals function is quadratic and symmetric around zero; the sum of absolute residuals is piecewise linear and symmetric around zero. Therefore, minimizing the sum of absolute residuals equates the number of positive and negative residuals and defines a plane (a line in the case of a simple regression specification) that "goes through" the median.

What if we wished to estimate the parameters of the regression line that correspond to quantiles other than the median? Koenker and Bassett (1978) and Bassett and Koenker (1982) showed that if $\widehat{\boldsymbol{\beta}}_\tau$ minimizes

$$\sum_{i=1}^{N} \rho_\tau \left(y_i - \mathbf{x}_i' \boldsymbol{\beta}_\tau \right) \tag{9.1}$$

where $\rho_\tau(u) = \{\tau - \mathbf{1}(u < 0)\}$. Therefore, $\widehat{\boldsymbol{\beta}}_\tau$ is the solution to the τth quantile regression. In other words, the solution produces the best-fitting plane that goes through the τth quantile of u. To fix ideas, suppose that $\tau = 1/2$. Then

$$\rho_\tau(u) = \begin{cases} \frac{1}{2}(u), & \text{if } u > 0 \\ -\frac{1}{2}(u), & \text{if } u < 0 \end{cases}$$

and minimizing (9.1) produces estimates corresponding to the median regression.

Quantile regressions for a given conditional quantile have two appealing properties relative to standard least-squares regressions. First, quantile regression estimates are extremely robust to outliers. The reason is quite simple. Consider the case of the median regression. If a large (positive or negative) value of u_i changes by a bit, the median value of u_i is unaffected, so the quantile regression estimates remain unchanged. Second, quantile regressions are equivariant to monotone transformations of the outcome variable y_i. This means that not only the regression of y_i goes through the median of y_i but also the regression of any $h(y_i)$ will go through the median of y_i for any monotone function, $h(\cdot)$. This is because the median is an order statistic that is invariant to such transformation.

Our reason for describing quantile regressions focuses on yet another appealing property. We estimate quantile regressions at various values of τ to understand how the effects of covariates vary across the conditional quantiles of the outcome. Koenker and Hallock (2001) and Drukker (2016) give informative introductions to quantile regressions and how they allow researchers to explore heterogeneity of effects.

9.2.1 MEPS examples

To demonstrate the value of quantile regressions, we use the 2004 Medical Expenditure Panel Survey (MEPS) data introduced in chapter 3 to estimate the effect of a change quantiles of errors. Note that the quantiles on the horizontal axis refer to quantiles of errors—not to quantiles of the outcome, exp_tot. Although it is tempting to interpret the effects as if they were applicable to observed quantiles of the outcome, that interpretation is incorrect. These are conditional quantiles, so they cannot be easily translated into unconditional quantiles.

Nevertheless, it is revealing that—in both panels—the quantile regression estimates lie outside the confidence intervals of the least-squares estimates for most quantiles, suggesting that the effects of these covariates are not constant across the error distribution or equivalently across the conditional distribution of the dependent variable. The OLS coefficient on age statistically coincides with quantile estimates from the 70th through 80th percentiles, while the OLS coefficient on female statistically coincides with quantile estimates from about the 35th through 80th percentiles. A researcher might conclude that there is more heterogeneity in the effect of a change in age than there is in the effect of gender on total expenditures.

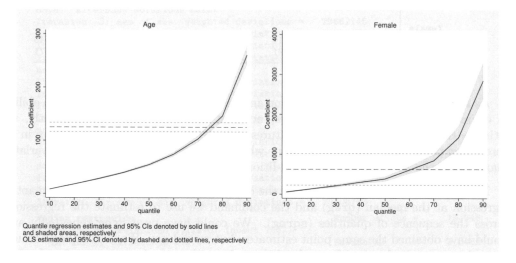

Figure 9.1. Coefficients and 95% confidence intervals by quantile of expenditure errors

As we described in chapter 3, the distribution of expenditures—conditional on being positive—is severely skewed to the right. In this case, the median quantile regression produces substantially different estimates than the least-squares estimates (mean regression).

What might we learn if the distribution of the outcome were more symmetric? To explore this, we estimate quantile regressions for the logarithm of total expenditures (conditional on expenditures being positive). We first estimate the quantile regression at the median using qreg. The results show that the coefficient on age is 0.037, implying that if an individual is one year older, that individual would spend 3.7% more. Women spend 43% {exp(0.36) − 1} more than men.

```
. *** Generate ln(total expenditures)
. generate ln_exp_tot = ln(exp_tot)

. *** Quantile (median) regression of ln(expenditures)
. qreg ln_exp_tot age i.female, quantile(0.5) vce(robust)
Iteration  1:  WLS sum of weighted deviations =    9523.677

Iteration  1: sum of abs. weighted deviations =   9523.6702
Iteration  2: sum of abs. weighted deviations =    9523.575
Iteration  3: sum of abs. weighted deviations =    9522.912
Iteration  4: sum of abs. weighted deviations =   9521.5422
note:  alternate solutions exist
Iteration  5: sum of abs. weighted deviations =   9520.0523
Iteration  6: sum of abs. weighted deviations =   9519.8278
note:  alternate solutions exist
Iteration  7: sum of abs. weighted deviations =   9519.7731
Iteration  8: sum of abs. weighted deviations =   9519.7437
Iteration  9: sum of abs. weighted deviations =   9519.7397
note:  alternate solutions exist
Iteration 10: sum of abs. weighted deviations =   9519.7295
Iteration 11: sum of abs. weighted deviations =   9519.7177
note:  alternate solutions exist
Iteration 12: sum of abs. weighted deviations =   9519.7175
Iteration 13: sum of abs. weighted deviations =    9519.717
```

```
Median regression                              Number of obs =      15,946
  Raw sum of deviations 10527.16 (about 7.3375878)
  Min sum of deviations 9519.717                Pseudo R2     =      0.0957
```

ln_exp_tot	Coef.	Robust Std. Err.	t	P>|t|	[95% Conf. Interval]	
age	.0371921	.000829	44.86	0.000	.0355671	.038817
female						
Female	.3608878	.0319231	11.30	0.000	.2983148	.4234607
_cons	5.292287	.050472	104.86	0.000	5.193356	5.391218

We use `sqreg` to display the effects of covariates across the quantiles of the conditional distribution of the logarithm of expenditures and to compare them with OLS estimates in figure 9.2. We find that there is still substantial evidence of effect heterogeneity for `age`. The pattern of coefficients is reversed, relative to the effects of `age` on expenditures as shown in figure 9.1. In the previous case, the effect of a change in age increased as the conditional quantile of expenditures increased. Now, when the outcome is the log of expenditure, the effect of age decreases across those conditional quantiles. There is no evidence of heterogeneity in the effect of `female` for expenditures measured on the log scale. The quantile estimates are all within the confidence interval of the OLS estimates.

Suppose that an outcome, y, is a random variable drawn from one of C distributions. Assume the probability that y is drawn from distribution (more commonly referred to as component or class) c is π_c, with $0 < \pi_c < 1$, and $\sum_{c=1}^{C} \pi_c = 1$. Let $f_c(y|\mathbf{x}; \boldsymbol{\theta_c})$ denote the density (mass function if y is a discrete random variable) for class or component c where $c = 1, 2, \ldots, C$ and $\boldsymbol{\theta_c}$ denote the parameters of the distribution, $f_c(\cdot)$. Then, the density function for a C-component finite mixture (Lindsay 1995; Deb and Trivedi 1997; McLachlan and Peel 2000) is the weighted sum of component densities,

$$f(y|\mathbf{x}; \boldsymbol{\theta_1}, \boldsymbol{\theta_2}, \ldots, \boldsymbol{\theta_C}; \pi_1, \pi_2, \ldots, \pi_C) = \sum_{c=1}^{C} \pi_c f_c(y|\mathbf{x}; \boldsymbol{\theta_c}) \tag{9.2}$$

This model's parameters can be estimated by maximum likelihood, either by an expectation-maximization (EM) algorithm (McLachlan and Peel 2000) or more standard Newton–Raphson methods (Deb and Trivedi 1997), or a combination of techniques as implemented in Stata. It is typical, in most estimation algorithms, not to estimate the π_c parameters directly. Instead, they are reparameterized as

$$\pi_c = \frac{\exp(\gamma_c)}{\sum_{k=1}^{C} \exp(\gamma_k)} \tag{9.3}$$

with $\gamma_1 = 0$ by convention as a normalization restriction. A major computational advantage of estimating the γ_c parameters is that there are no restrictions on their values unlike the values of π_c each of which is constrained to be between zero and one and they must sum to one.

Arguably, the most common component density is the normal (Gaussian) distribution. Its functional forms and properties have been described in numerous works, including McLachlan and Peel (2000). Other popular choices of distributions in the literature for continuous outcomes are the lognormal and generalized linear models (GLMs). Popular choices for distributions of count outcomes are the Poisson and negative binomial. Generally, the finite mixture model can be specified with any density the researcher deems appropriate for the outcome.

To illustrate the model in the case of an important distribution choice in the context of healthcare expenditures, we describe specific details of the finite mixture of a generalized linear regression with a gamma density and log link. Recall that in chapter 5, we showed that the GLM with a gamma density and log link described expenditures well. For individuals in class c, the GLM with a gamma density and log-link function for an outcome y can be expressed as a density function (McCullagh and Nelder 1989) using

$$f_c(y_i|\mathbf{x}_i; \boldsymbol{\theta_c}) = -\frac{y_i}{\exp\{(\mathbf{x}_i'\boldsymbol{\beta}_c)\}} + \frac{1}{\exp \mathbf{x}_i'\boldsymbol{\beta}_c} \tag{9.4}$$

The mixture density in this model is obtained by replacing $f_c(y_i|\mathbf{x}_i; \boldsymbol{\theta_c})$ from (9.4) into (9.2).

In this parameterization, the expected value of the outcome, y_i, given covariates $\mathbf{x}_i$ and class c, is

$$E_c(y_i|\mathbf{x}_i) = \exp(\mathbf{x}_i'\boldsymbol{\beta}_c) \tag{9.5}$$

Consequently, using properties of finite mixture distributions, we see that the expected value of the outcome, y_i, given covariates $\mathbf{x}_i$, is

$$E(y_i|\mathbf{x}_i) = \sum_{c=1}^{C} \pi_c \left\{ \exp\left(\mathbf{x}_i'\boldsymbol{\beta}_c\right) \right\}$$

The expected value formula in (9.5) can be used to estimate heterogeneous effects of treatment, as well as marginal and incremental effects.

In the finite mixture models described above, the π_c parameters can be interpreted as prior probabilities of class membership. As specified, they are assumed to be constants. Therefore, they do not provide any information on how likely a particular observation might be to belong to a particular class c. However, following the finite mixture literature, we can calculate the posterior probability that observation i belongs to component c:

$$\Pr\left(i \in \text{class } c|\mathbf{x}_i, y_i; \boldsymbol{\theta}\right) = \frac{\pi_c f_c(y_i|\mathbf{x}_i, \boldsymbol{\theta}_c)}{\sum\limits_{k=1}^{C} \pi_k f_k(y_i|\mathbf{x}_i, \boldsymbol{\theta}_k)}, \qquad c = 1, 2, \ldots C \qquad (9.6)$$

These posterior probabilities vary across individuals and provide a mechanism for assigning individuals to latent classes and for using those assignments to characterize the classes or components. To do so, we would estimate the posterior probabilities of class membership after the model parameters have been estimated. Then, we could use the probabilities themselves—or classification based on the probabilities—to further describe characteristics of observations in each class. Note that although it is technically possible to parameterize the prior probabilities to allow them to vary by characteristics, the tradition in the literature is to assume they are constant (McLachlan and Peel 2000).

The discussion above has taken the number of classes, C, as given. In practice, the number of classes is not likely to be known a priori, so the researcher will need to determine C empirically. There is a subtle identification issue that complicates testing the number of components in a finite mixture model using standard test statistics (see Lindsay [1995] and McLachlan and Peel [2000] for discussions). Instead, the literature points to the Akaike information criterion (AIC) and Bayesian information criterion (BIC) as consistent tools for model selection (Leroux 1992). In practice, a two-component model is fit first and the model-selection criteria computed. Then models with additional components are fit until model-selection criteria suggest no improvement.

Below we describe two empirical examples. The first estimates finite mixtures of GLMs with gamma densities and log links for healthcare expenditures. The second estimates finite mixtures of negative binomial regressions for counts of office-based visits. Finite mixture models for count data, and details of finite mixture models more generally, are described in Deb and Trivedi (1997, 2002).

If we thought this model was adequate, we would proceed to interpreting estimates and characterizing marginal effects. However, as we will see below, it is not. Nevertheless, to fix ideas using this simple case, we briefly interpret the parameter estimates. The output of parameter estimates is shown in three blocks. The first block displays estimates of the γ_c parameters, which can be used to calculate the class probabilities using (9.3). We do this below, but after interpreting the output from the next two blocks of results that show the componentwise parameter estimates. For observations in component 1, the effects of `age` and `female` are both statistically significant—each increases spending. For observations in component 2, `age` is statistically significant, but `female` is not.

Now we return to calculating the latent class probabilities. We use the postestimation command `estat lcprob` to obtain estimates of π_c. The two components have probabilities, π_1 and π_2, equal to 0.68 and 0.32, respectively.

```
. *** Two-component FMM class probabilities: Total expenditures
. estat lcprob

Latent class marginal probabilities               Number of obs    =    15,946
```

	Margin	Delta-method Std. Err.	[95% Conf. Interval]	
Class				
1	.6827209	.0236649	.634641	.7271935
2	.3172791	.0236649	.2728065	.365359

Having described results from a two-component finite mixture of gamma GLMs using `fmm`, we fit a three-component model that may be a better description of the data-generating process. The output from estimation of a three-component finite mixture model using `fmm` produces four blocks of iteration logs and, this time, four blocks of parameter estimates. Despite all the effort to obtain good starting values, it is not surprising to see a number of iterations labeled (`not concave`) prior to convergence—as in the output below. The first block of parameter estimates presents estimates for the latent class probabilities. The next three blocks show results for the three componentwise regressions. Except for the coefficient on `female` in component 3, all other covariate coefficients are statistically significant. Older individuals spend more in each latent class. Women spend more than men in the first and second classes. We will return to a more detailed interpretation of the estimates using `margins` below.

```
. *** Three-component FMM parameter estimates: Total expenditures
. fmm 3,  vce(robust): glm exp_tot age i.female, family(gamma) link(log)
Fitting class model:

Iteration 0:   (class) log likelihood = -17518.471
Iteration 1:   (class) log likelihood = -17518.471

Fitting outcome model:

Iteration 0:   (outcome) log likelihood =  -143395.1
Iteration 1:   (outcome) log likelihood = -142102.56
Iteration 2:   (outcome) log likelihood = -142076.88
Iteration 3:   (outcome) log likelihood = -142076.83
Iteration 4:   (outcome) log likelihood = -142076.83

Refining starting values:

Iteration 0:   (EM) log likelihood = -158443.72
Iteration 1:   (EM) log likelihood = -157974.05
Iteration 2:   (EM) log likelihood = -157844.22
Iteration 3:   (EM) log likelihood = -157812.72
Iteration 4:   (EM) log likelihood = -157798.15
Iteration 5:   (EM) log likelihood = -157782.27
Iteration 6:   (EM) log likelihood = -157762.51
Iteration 7:   (EM) log likelihood = -157738.61
Iteration 8:   (EM) log likelihood = -157710.22
Iteration 9:   (EM) log likelihood =  -157677.3
Iteration 10:  (EM) log likelihood = -157640.08
Iteration 11:  (EM) log likelihood = -157599.05
Iteration 12:  (EM) log likelihood =  -157554.8
Iteration 13:  (EM) log likelihood = -157507.91
Iteration 14:  (EM) log likelihood = -157458.99
Iteration 15:  (EM) log likelihood = -157408.56
Iteration 16:  (EM) log likelihood = -157357.13
Iteration 17:  (EM) log likelihood = -157305.14
Iteration 18:  (EM) log likelihood = -157253.02
Iteration 19:  (EM) log likelihood = -157201.11
Iteration 20:  (EM) log likelihood = -157149.73
Note: EM algorithm reached maximum iterations.

Fitting full model:

Iteration 0:   log pseudolikelihood =  -144548.6  (not concave)
Iteration 1:   log pseudolikelihood = -144532.58  (not concave)
Iteration 2:   log pseudolikelihood = -144501.88
Iteration 3:   log pseudolikelihood = -144468.98
Iteration 4:   log pseudolikelihood =  -144459.8
Iteration 5:   log pseudolikelihood = -144459.39
Iteration 6:   log pseudolikelihood = -144459.39
```

Finite mixture model   Number of obs   =   15,946
Log pseudolikelihood = -144459.39

	Coef.	Robust Std. Err.	z	P>|z|	[95% Conf. Interval]	
1.Class	(base outcome)					
2.Class						
_cons	-.1238305	.0913192	-1.36	0.175	-.3028128	.0551518
3.Class						
_cons	-1.815168	.2581631	-7.03	0.000	-2.321158	-1.309178

```
. *** FMM AIC and BIC: Total expenditures
. estimates stats fmm2 fmm3
Akaike's information criterion and Bayesian information criterion
```

Model	Obs	ll(null)	ll(model)	df	AIC	BIC
fmm2	15,946	.	-144712.6	9	289443.2	289512.3
fmm3	15,946	.	-144459.4	14	288946.8	289054.3

Note: N=Obs used in calculating BIC; see [R] BIC note.

As a sanity check, we estimate the overall predicted mean using `margins` and its empirical counterpart for the sample of observations with positive values of `exp_tot`. It is an indicator of model validity that the predicted mean is quite close to the empirical estimate.

```
. *** Three-component FMM predicted and empirical means: Total expenditures
. margins
Predictive margins                          Number of obs    =      15,946
Model VCE    : Robust

Expression   : Predicted mean (Total medical care expenses), using class
               probabilities, predict(mu outcome(exp_tot))
```

	Margin	Delta-method Std. Err.	z	P>|z|	[95% Conf. Interval]	
_cons	4498.545	82.30459	54.66	0.000	4337.231	4659.859

```
. summarize exp_tot
```

Variable	Obs	Mean	Std. Dev.	Min	Max
exp_tot	15,946	4480.262	10604.14	2	440524

We begin the characterization of results from the three-component model by using `margins` to compute means of predicted outcomes for each of the three components. Using the results below, we can say that the distribution of expenditures is characterized by three components, one with mean spending of \$1,322 and associated probability of 0.49, a second component with mean spending of \$5,245 and probability of 0.43, and a third, relatively rare class (with probability 0.08) with mean spending of \$19,959. We can now say that, while women spend more than men in the low- and medium-spending classes, women and men in the high-spending class, that is, in component 3, do not differ significantly.

```
. *** Three-component FMM component-level predicted means: Total expenditures
. estat lcmean
Latent class marginal means                    Number of obs      =     15,946
```

		Margin	Delta-method Std. Err.	z	P>|z|	[95% Conf. Interval]	
1	exp_tot	1322.05	69.82991	18.93	0.000	1185.185	1458.914
2	exp_tot	5244.849	419.6181	12.50	0.000	4422.413	6067.285
3	exp_tot	19958.93	2122.849	9.40	0.000	15798.22	24119.63

It is not just mean spending that differs across components. The shapes of the gamma densities are also different. To show this, we plot the predicted densities at the median **age** and **gender** in figure 9.3. To make the figure easier to read, we show only the densities through $20,000 in expenditure, which exceeds the 95th percentile of its empirical distribution of expenditure. The figure shows that each of the predicted densities is skewed, just as the empirical density is. But, while the densities of the first two components have positive modes (classically gamma shaped), the density of component 3 is exponential in shape—it slopes downward right from the beginning.

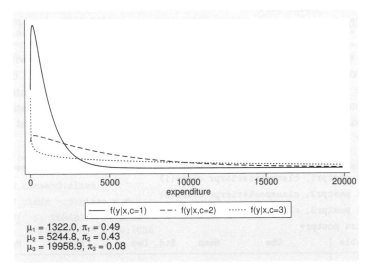

Figure 9.3. Empirical and predicted componentwise densities of expenditure

What do the parameter estimates of this three-component model tell us about the effects of **age** and **female** on expenditures? We present the estimates of the componentwise partial effects, along with their 95% confidence intervals graphically. In the figure,

```
. regress female i.class

      Source │       SS           df       MS            Number of obs   =     15,946
─────────────┼──────────────────────────────────        F(2, 15943)     =       7.98
       Model │  3.85848471         2   1.92924236        Prob > F        =     0.0003
    Residual │  3852.06024     15,943   .241614517       R-squared       =     0.0010
─────────────┼──────────────────────────────────        Adj R-squared   =     0.0009
       Total │  3855.91873     15,945   .241826198       Root MSE        =     .49154


      female │      Coef.   Std. Err.      t    P>|t|     [95% Conf. Interval]
─────────────┼────────────────────────────────────────────────────────────────
       class │
          2  │   -.0026953   .0081446    -0.33   0.741    -.0186595     .013269
          3  │   -.0877488   .0219967    -3.99   0.000    -.1308648    -.0446328
             │
       _cons │    .5943902   .0050381   117.98   0.000     .5845149     .6042654

. test 2.class=3.class

 ( 1)  2.class - 3.class = 0

       F(  1, 15943) =     14.48
            Prob > F =    0.0001
```

9.3.2   MEPS example of healthcare use

In a second example, we fit finite mixture models for the number of office-based health-care visits. We showed in chapter 8 that the negative binomial-1 fit this outcome well. Therefore, for this example, we estimate finite mixtures of negative binomial-1 regressions. We begin by fitting a two-component model—but for brevity, we do not show the results. Once again, information criteria suggest that the three-component model is better than the two-component one.

When we first fit the three-component model, we noticed that it took 55 iterations to converge (after 20 iterations of the EM algorithm), including 49 iterations showing the dreaded not concave message. While there is nothing wrong with such an iterative process, especially in the context of finite mixture models, we were impatient, so we used the difficult option for optimization. With this option, the final maximum-likelihood estimation algorithm converges in seven iterations. From our experience, we note that difficult may be a useful option in many circumstances.

Estimates from the three-component model are shown below. The coefficients on age and female are statistically significant in each of the three components.

```
. *** Three-component FMM parameter estimates: # office-based visits
. fmm 3, vce(robust) difficult: nbreg use_off age i.female, dispersion(constant)

Fitting class model:

Iteration 0:   (class) log likelihood = -21297.698
Iteration 1:   (class) log likelihood = -21297.698

Fitting outcome model:

Iteration 0:   (outcome) log likelihood = -54249.857
Iteration 1:   (outcome) log likelihood = -51757.547
Iteration 2:   (outcome) log likelihood = -47160.415
Iteration 3:   (outcome) log likelihood = -46124.367
Iteration 4:   (outcome) log likelihood = -45426.015
Iteration 5:   (outcome) log likelihood = -45410.444
Iteration 6:   (outcome) log likelihood = -45410.391
Iteration 7:   (outcome) log likelihood = -45410.391

Refining starting values:

Iteration 0:   (EM) log likelihood = -64739.188
Iteration 1:   (EM) log likelihood = -63690.327
Iteration 2:   (EM) log likelihood = -63646.677
Iteration 3:   (EM) log likelihood = -64035.313
Iteration 4:   (EM) log likelihood = -64489.275
Iteration 5:   (EM) log likelihood = -64900.348
Iteration 6:   (EM) log likelihood = -65249.473
Iteration 7:   (EM) log likelihood = -65541.111
Iteration 8:   (EM) log likelihood =  -65784.22
Iteration 9:   (EM) log likelihood = -65987.142
Iteration 10:  (EM) log likelihood = -66156.693
Iteration 11:  (EM) log likelihood = -66298.268
Iteration 12:  (EM) log likelihood = -66416.173
Iteration 13:  (EM) log likelihood = -66513.931
Iteration 14:  (EM) log likelihood = -66594.501
Iteration 15:  (EM) log likelihood = -66660.423
Iteration 16:  (EM) log likelihood = -66713.912
Iteration 17:  (EM) log likelihood = -66756.907
Iteration 18:  (EM) log likelihood = -66791.108
Iteration 19:  (EM) log likelihood = -66817.993
Iteration 20:  (EM) log likelihood = -66838.832
Note: EM algorithm reached maximum iterations.

Fitting full model:

Iteration 0:   log pseudolikelihood = -50350.841  (not concave)
Iteration 1:   log pseudolikelihood = -50336.985
Iteration 2:   log pseudolikelihood = -50334.611
Iteration 3:   log pseudolikelihood = -50332.851
Iteration 4:   log pseudolikelihood = -50331.731
Iteration 5:   log pseudolikelihood = -50331.526
Iteration 6:   log pseudolikelihood =   -50331.5
Iteration 7:   log pseudolikelihood = -50331.499
```

```
. *** Three-component FMM component-level predicted means: # office-based visits
. estat lcmean

Latent class marginal means                        Number of obs     =     19,386
```

		Margin	Delta-method Std. Err.	z	P>|z|	[95% Conf. Interval]	
1							
	use_off	3.482261	.1604877	21.70	0.000	3.167711	3.796811
2							
	use_off	3.047602	.399561	7.63	0.000	2.264477	3.830727
3							
	use_off	12.11311	.8076261	15.00	0.000	10.53019	13.69603

From the estimates of the predicted means, one might conclude that two of the components are too similar to distinguish. That conclusion would be wrong; the densities of the components are substantially different from each other. To demonstrate this, we plot the predicted densities at the median age and gender in figure 9.5. To make the figure easier to read, we show only the densities through 30 visits, which exceeds the 97th percentile of its empirical distribution of office-based visits. We also represent the densities using (continuous) line charts, although bar charts would be technically preferred. We use line charts because it is easier to visualize differences in the component densities. The figure shows that each of the predicted densities is skewed, just like the empirical density. The density of the relatively rare component 2 is quite different from that of the much more frequent component 1, although they have similar means. Observations in component 3 are most likely to generate large values of visits and much less likely to generate zero and other small-visit values.

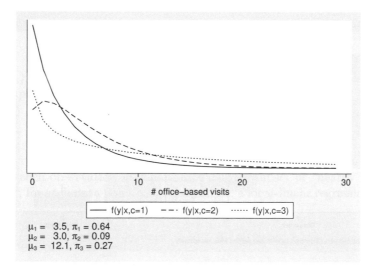

Figure 9.5. Empirical and predicted componentwise densities of office-based visits

We take the estimated marginal effects of `age`, and the incremental effects of `female`, and display them along with 95% confidence intervals in figure 9.6 below. The left panel displays marginal effects of `age`. The estimated marginal effect of `age` is negative and statistically significant for the lowest use (on average) group. The effects of age are positive for the middle-use group—and positive and substantially bigger for the high-use group. Both effects are statistically significant. The effects of `female` are distinctly nonmonotonic across the components, ordered in ascending values of mean predicted use. Among the 9% of individuals in the component with the lowest office-based use, the effect of `female` is large and positive, albeit with a wide confidence interval. Among the 64% of individuals in the component with moderate office-based use, the effect of gender is positive and statistically significant but quite small. Among the 27% of individuals in the high-use component, the effect of `female` is moderately large and statistically significant.

```
. *** Restrict to subsample with positive total expenditures
. *** Restrict to sample of Hispanic males
. drop if exp_tot <= 0
(3,440 observations deleted)
. keep if female==0 & eth_hisp==1
(14,876 observations deleted)
. regress exp_tot age pcs i.anylim
```

Source	SS	df	MS		
Model	6.1455e+09	3	2.0485e+09	Number of obs	= 1,070
Residual	3.6476e+10	1,066	34217245.8	F(3, 1066) = 59.87
				Prob > F = 0.0000
				R-squared = 0.1442
				Adj R-squared = 0.1418
Total	4.2621e+10	1,069	39870047.3	Root MSE = 5849.6

exp_tot	Coef.	Std. Err.	t	P>|t|	[95% Conf. Interval]
age	50.75679	12.10509	4.19	0.000	27.00427 74.5093
pcs12	-124.706	20.30998	-6.14	0.000	-164.5581 -84.85394
anylim					
Activity ..	2246.889	473.6344	4.74	0.000	1317.528 3176.251
_cons	5820.033	1284.59	4.53	0.000	3299.421 8340.644

Next, we use `npregress` to estimate the effects nonparametrically. We use a nonparametric bootstrap (with 100 bootstrap replications) to obtain standard errors, test statistics, and confidence intervals, because `npregress` does not produce those by default. See Cattaneo and Jansson (2017) for formal justification of the bootstrap for the nonparametric regression. We have used 100 replications to economize on computational time without confirming that the number of replications is sufficient for reliable estimates. For serious research work, we encourage users to fit models with different numbers of bootstrap replicates before settling on a number beyond which estimates of standard errors do not change much.

The output of `npregress` shows the average of the predicted means and the averages of the predicted derivatives (changes for discrete regressors) of the mean function with bootstrap standard errors. The average of the observation-level effects of a activity limitation is $2,247. Individuals with disabling activity limitations have higher healthcare expenditures than those who do not. Note that this estimate is almost 45% larger than the OLS estimated effect of $1,554. The average of the effects of `pcs` is −$109, while its OLS counterpart is −$125. In both models, the interpretation of the effect is that better health (indicated by higher `pcs12` scores) is associated with lower healthcare expenditures. While the nonparametric average of the effects is not so different from the effect estimated by OLS, one should not conclude that OLS estimates are "good enough". There may well be important nonlinearities and interactions in the effects of the physical health score and activity limitations on expenditures that the nonparametric regression takes into account (but OLS does not). These nonlinearities may be of substantive interest in many applications.

```
. npregress kernel exp_tot age pcs i.anylim, reps(100)
(running npregress on estimation sample)

Bootstrap replications (100)
———+— 1 ——+— 2 ——+— 3 ——+— 4 ——+— 5
.................................................      50
.................................................     100
```

Bandwidth

	Mean	Effect
Mean		
age	5.890194	26.3865
pcs12	3.839381	17.19941
anylim	.5	.5

```
Local-linear regression               Number of obs    =       1,069
Continuous kernel : epanechnikov       E(Kernel obs)    =       1,069
Discrete kernel   : liracine           R-squared        =      0.2426
Bandwidth         : cross validation
```

exp_tot	Observed Estimate	Bootstrap Std. Err.	z	P>|z|	Percentile [95% Conf. Interval]	
Mean						
exp_tot	2545.919	221.3198	11.50	0.000	2058.948	2944.794
Effect						
age	45.57655	10.25433	4.44	0.000	25.63205	65.61113
pcs12	-108.7793	19.30334	-5.64	0.000	-149.1225	-74.26577
anylim (Activity.. vs No activ..)	1554.459	420.7119	3.69	0.000	682.1224	2294.557

```
Note: Effect estimates are averages of derivatives for continuous covariates
      and averages of contrasts for factor covariates.
```

To understand the effect of a change in a unit of pcs12 better, we use margins to compute the conditional mean function at values across the empirical distribution of pcs12 and anylim. Once again, we use 100 bootstrap replications to obtain standard errors for the predictions.

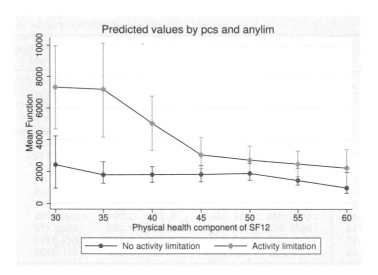

Figure 9.7. Predicted total expenditures by physical health score and activity limitation

We conclude our description of the nonparametric regression estimator with a report on computational times because they can be considerable. It took about 2 minutes to obtain estimates of the average of effects and their standard errors based on 100 bootstrap replications in our example with 3 regressors and just over 1,000 observations. It took about 15 minutes to obtain the output of `margins` with standard errors obtained by bootstrap. Each call to `margins` takes about the same amount of time as the regression estimates; we evaluated `margins` at 14 values. These are quite modest times. When we estimated the same specification using a different sample of about 8,000 observations, the regression estimates (with standard errors calculated using 100 bootstrap replications) took about 1.5 hours. It took about 9 hours to obtain the results of `margins` with 100 bootstrap replications. These calculations were done using a 4-core version of Stata-MP on a reasonably fast Linux machine. This is the cost of an estimator that requires almost no judgment on the part of the researcher in terms of functional forms of covariates and relationships between them. In some situations, the extra effort may well be worth it. In other situations, researchers may find that getting graphs of effects without confidence intervals (which is the source of most of the computational time) may be sufficiently insightful.

9.5   Conditional density estimator

We end this chapter with a brief description of another flexible estimator proposed by Gilleskie and Mroz (2004). Their conditional density estimator (CDE) is easy to estimate. The intuition is to break up the domain of the dependent variable into a set of mutually exclusive and exhaustive subdomains—which we will call bins—and focus the modeling effort on predicting the probability of being in each bin. The mean value of the dependent variable is assumed constant within each bin. This assumption is reasonable when bin sizes are small. After estimation, the results can estimate predicted means conditional on the covariates, as well as marginal and incremental effects.

The primary advantage of CDE is that it can be used with continuous and count outcomes. It works with single- and multipeaked distributions. For example, it can model the number of hours worked per week, where there are typically peaks in the density at 20, 35, and 40 hours. CDE can also model medical care expenditure data, incorporating zeros easily. CDE can be extended to count models, for example, to model the number of prescription drugs purchased in a year (Mroz 2012). Gilleskie and Mroz (2004) built on earlier work by Efron (1988) and Donald, Green, and Paarsch (2000).

The two key assumptions in CDE (see Gilleskie and Mroz [2004], equations 12 and 13) are that the probability of being in a bin depends on covariates and that the mean value of y, conditional on the bin, is independent of the covariates. That is, there is heterogeneity across bins but homogeneity within bins. Within a bin, covariates have no predictive power. In this case, the CDE approach focuses on modeling the probability of being in a bin in the best possible way.

The second assumption greatly simplifies estimation and inference, because predicted expenditures are the same for all within the same bin. The assumption of homogeneity within a bin is typically least likely to hold for the highest expenditure group. In principle, the second assumption could be relaxed by treating the last bin like the second part of a two-part model.

Given the two main assumptions, fitting the CDE model is straightforward. Decide on the number of bins and the threshold values separating the bins, which do not need to be equally spaced. Fit a series of logit (or probit) models. For example, if there are 11 bins, fit 10 logit models, with each successive model fit on a smaller sample. For observations $i = 1, \ldots, N$ and bins $k = 1, \ldots, K$ the expected value of the dependent variable, y_i, is the mean of y_i for each bin times the probability of being in that bin, summed over all bins:

$$E\left(y_i|\mathbf{x}_i\right) = \sum_{k=1}^{K} \Pr\left\{y_i \text{ in bin } k|\mathbf{x}_i, y_i \text{ not in bins 1 to } (k-1)\right\}$$
$$\times E\left(y_i|\mathbf{x}_i, y_i \text{ in bin } k\right)$$

Gilleskie and Mroz (2004) demonstrate how to fit CDE in one large logit model, as opposed to a series of individual logit models with progressively smaller sample sizes.

Two issues of this model have not yet been worked out to make this method accessible to the typical applied researcher. First, it needs a theory and practical method for choosing the number of bins in an optimal way. Second, although methods for the computation of standard errors are available, coding those is beyond the scope of this book.

9.6   Stata resources

The main quantile command in Stata is `qreg`. To estimate a sequence of quantile regressions, as demonstrated in section 9.2.1, use `sqreg`.

A new Stata command, `fmm`, fits a variety of finite mixture models. In prior versions of Stata, finite mixture models could be fit with a user-written package (Deb 2007) with the same name. With the introduction of the new `fmm` command, `fmm` points to the official Stata command, even if the user has the package already installed. In older versions of Stata, to run finite mixture models, use the package `fmm9`. Install this package by typing `ssc install fmm9` (Deb 2007).

Nonparametric regressions can be estimated in Stata using `npregress`. More specifically, `npregress` estimates local linear and local constant regressions to produce observation-level effects of covariates on an outcome.

10 Endogeneity

10.1 Introduction

The models in this book specify how a dependent variable is generated from exogenous, observed covariates and unobserved errors. A covariate that is uncorrelated with the unobserved errors is exogenous. In contrast, a covariate that is correlated with the unobserved errors is endogenous. Endogeneity is a problem because correlation between a covariate and the error term will cause inconsistency in the estimated coefficients in the models we have described so far. In the parlance of chapter 2, the ignorability assumption does not hold, and estimates of treatment effects will not be consistent if we use models that ignore the endogeneity.

One common cause of endogeneity in health economics is unobserved health status. Consider a model that predicts healthcare expenditures as a function of out-of-pocket price. Even after controlling for observed health status, unobserved health status remains in the error term because health status is hard to measure accurately and fully. Unobserved health status is correlated with both out-of-pocket price and healthcare expenditures. Therefore, an ordinary least-squares (OLS) regression (or any of the other methods for estimating model parameters described in previous chapters) of healthcare expenditures on out-of-pocket price will produce inconsistent estimates of the coefficient on out-of-pocket price. In some literature, these omitted variables are called confounders.

In this chapter, we provide a brief introduction to the issues raised by endogeneity and a few available solutions. Many econometric methods to control for endogeneity use variables, called instrumental variables (IVs), that predict the endogenous variable but do not directly predict the main dependent variable. We use simulated data to show how to use IVs to control for endogeneity. We begin our description with a discussion of two-stage least squares (2SLS) to solve the problem of inconsistency because of endogeneity. In this context, we discuss standard statistical tests for endogeneity and for instrument validity. The control function approach is an important 2SLS extension to deal with endogeneity in models in which either the outcome or the endogenous regressor is not continuous, thus suggesting the use of nonlinear models. A common version of control functions is two-stage residual inclusion (2SRI). We also describe Stata's extended regression models (ERM), a unified framework for dealing with endogeneity in linear and some nonlinear models. This framework allows for endogenous and exogenous covariates and will estimate treatment effects under the assumption of jointly normal errors. Finally, we also briefly describe the generalized method of moments (GMM) for IVs estimation, because GMM can have substantial benefits compared with 2SLS.

This chapter is not a comprehensive overview of econometric issues in endogeneity, measurement error, and use of IVs. For further information on models with endogeneity, we direct readers to the general econometric literature. There are excellent summaries on this matter in recent econometric textbooks by Cameron and Trivedi (2005), Wooldridge (2010), and Angrist and Pischke (2009) and in review articles by Newhouse and McClellan (1998), Angrist and Krueger (2001), and Murray (2006).

10.2   Endogeneity in linear models

10.2.1   OLS is inconsistent

We use a simple, canonical example to help clarify how endogeneity can create inconsistent coefficient estimates. The outcome of interest, y_1, is continuous and determined by a linear equation

$$y_1 = \beta_0 + y_2\beta_1 + x\beta_2 + u + \varepsilon_1 \tag{10.1}$$

where x and y_2 are observed characteristics (covariates). Think of u as being an unobserved covariate and ε_1 as a typical linear regression error term. The β_j are parameters to be estimated. The covariate x is exogenous; it is uncorrelated with both u and ε_1. The covariate y_2 is determined by a linear equation

$$y_2 = \gamma_0 + w\gamma_1 + x\gamma_2 + u + \varepsilon_2 \tag{10.2}$$

where w is an exogenous covariate and ε_2 is the regression error term. Note that the unobserved term, u, enters both regression equations. The reason that y_2 is endogenous in the model defined in (10.1) is that it is a function of unobserved u and therefore correlated with the composite error $(u + \varepsilon_1)$. The model defined in (10.1) violates the assumption that the errors are uncorrelated with observed characteristics, so OLS estimates of its parameters will not be consistent. For example, let y_1 be healthcare spending and y_2 be out-of-pocket price. If both spending and out-of-pocket price depend on observed characteristics such as age, education (x), and unobserved health status (u), then the OLS estimates of the effect of out-of-pocket price on spending will be inconsistent.

Note that we have not yet described the purpose of including the covariate w in (10.2). We will do so in the next section, where we describe solutions to the problem described here. Note also that the logic described above applies even if ε_1 and ε_2 are uncorrelated with each other; the composite error terms, $(u + \varepsilon_1)$ and $(u + \varepsilon_2)$, are correlated, thus rendering OLS estimates of parameters of (10.1) inconsistent. In fact, one need not construct this example with an unobserved covariate u as distinct from the error terms ε_1 and ε_2. As long as the errors of (10.1) and (10.2) are correlated, OLS estimates of the model for the outcome, y_1, will be inconsistent.

We now use artificial data to demonstrate empirically how the omitted variable leads to inconsistency in OLS estimates. We generated a dataset with 10,000 observations. As shown below, we begin by drawing data for the observed covariates x and w from standard normal distributions. We also draw the unobserved variables, u, e1, and e2, from standard normal distributions. Each of these random variables is drawn independently. The endogenous covariate y2, following (10.2), is a function of x, u, the unobserved error e2, and a covariate w. Finally, we generate the dependent variable, y1, using (10.1) as a function of an exogenous covariate x, an endogenous covariate y2, the unobserved covariate u, and the unobserved error e1. Note that w enters the function for y2 but does not explicitly enter the function for y1. The variable w affects only y1 through its effect on y2. The unobserved component u affects both y1 and y2, which causes y2 to be endogenous. If we could observe u, then we could include u in the model for y1, and y2 would be exogenous instead of endogenous. For purposes of the example, we draw all the variables using a standard normal distribution random-number generator using the following commands:

```
. set obs 10000
number of observations (_N) was 0, now 10,000

. set seed 123456

. generate x  = rnormal(0,1)        // exogenous variable

. generate w  = rnormal(0,1)        // instrumental variable

. generate u  = rnormal(0,1)        // omitted variable

. generate e1 = rnormal(0,1)        // outcome eq error

. generate e2 = rnormal(0,1)        // endogenous regressor eq error

. generate y2 = x + .2*w + u + e2   // endogenous regressor equation

. generate y1 = y2 + x + u + e1     // outcome equation
```

When we fit the OLS model with u included as an observed covariate, the estimated coefficient on y2 is a consistent estimate of the true value of 1.0. We find that the estimated coefficient is close to 1.0.

```
. *** OLS: Regression with u included
. regress y1 y2 x u
```

Source	SS	df	MS		
Model	90316.5209	3	30105.507		
Residual	9995.42111	9,996	.999942088		
Total	100311.942	9,999	10.0321974		

				Number of obs	=	10,000
				F(3, 9996)	=	30107.25
				Prob > F	=	0.0000
				R-squared	=	0.9004
				Adj R-squared	=	0.9003
				Root MSE	=	.99997

y1	Coef.	Std. Err.	t	P>|t|	[95% Conf. Interval]	
y2	.988132	.009785	100.98	0.000	.9689515	1.007313
x	1.012624	.0139347	72.67	0.000	.9853087	1.039938
u	1.005259	.0139192	72.22	0.000	.9779749	1.032544
_cons	-.0041235	.0100005	-0.41	0.680	-.0237265	.0154795

Next, we omit u from the regression model. We know that the OLS estimate of the coefficient on y2 is inconsistent. In the example, the OLS estimate of the coefficient on y2 is 1.48 when u is omitted, and the confidence interval is far away from the true value of 1.0. The estimated coefficient when u is omitted is greater than 1.0, because the composite error terms of the two equations, (10.1) and (10.2), are positively correlated.

The second-stage regression (for the outcome of interest) predicts y1 as a function of the predicted endogenous variable, y2, and the exogenous variable, x. The 2SLS estimate is 0.932 with a standard error of 0.075. The estimate is close to 1 in magnitude, not statistically significantly different from 1.0 if one conducted a simple t test. However, it has a much wider confidence interval than found with OLS with no endogeneity. This example demonstrates two important points: First, with a valid instrument, the 2SLS estimate is much closer to the true value than the OLS estimate. Second, the standard errors are typically much larger than OLS. There is a tradeoff between consistency and precision.

10.2.3 Specification tests

As always, it is important to test both the validity of the instruments and whether endogeneity is in fact a problem. Stata's `ivregress` postestimation commands include the standard tests. The first set of tests are whether the instruments are strongly correlated with the potentially endogenous variable. In addition to inspecting t statistics on the instruments individually in the first-stage regression, you must inspect the F statistic on the joint test of significance of all the instruments. A joint test is important with multiple instruments because correlation among instruments—if there are more than one—can reduce individual t statistics and mask joint statistical significance. The rule of thumb is when there is one IV, the F statistic should be at least 10 (Nelson and Startz 1990; Staiger and Stock 1997; Stock, Wright, and Yogo 2002).

The `estat firststage` command reveals that the F statistic on the one instrument in the example is about 189, which is well above the minimum recommended threshold. We conclude that the instrument, w, strongly predicts the endogenous variable.

```
. *** Specification tests: First stage
. estat firststage
First-stage regression summary statistics
```

Variable	R-sq.	Adjusted R-sq.	Partial R-sq.	F(1,9997)	Prob > F
y2	0.3441	0.3440	0.0186	189.422	0.0000

```
Minimum eigenvalue statistic = 189.422

Critical Values                      # of endogenous regressors:     1
Ho: Instruments are weak             # of excluded instruments:      1
```

	5%	10%	20%	30%
2SLS relative bias		(not available)		

	10%	15%	20%	25%
2SLS Size of nominal 5% Wald test	16.38	8.96	6.66	5.53
LIML Size of nominal 5% Wald test	16.38	8.96	6.66	5.53

The second test is whether the potentially endogenous variable is actually exogenous.

This test is conditional on all the instruments being valid. The `estat endogenous` command shows that the p-value is below 0.05, meaning the test rejects the null hypothesis of exogeneity (conditional on the instrument being valid). We will treat `y2` as endogenous.

```
. *** Specification test: Exogeneity
. estat endogenous

  Tests of endogeneity
  Ho: variables are exogenous

  Durbin (score) chi2(1)        =   76.1288  (p = 0.0000)
  Wu-Hausman F(1,9996)          =   76.6822  (p = 0.0000)
```

In situations when there are more instruments than endogenous regressors, it is possible to test the overidentifying restrictions. That is, we could test the assumption that the additional instruments are unrelated to the main dependent variable, conditional on one of the instruments being valid. For a single instrument, one must rely on theory and an understanding of the institutions studied. In addition to these statistical tests, researchers also often perform balance tests to argue that values of covariates are balanced across different values of the instrument. If the instrument is purely random, it should be uncorrelated with the other exogenous variables (of course, it still needs to be correlated with the endogenous variable).

10.2.4 2SRI

There is another equivalent way to construct 2SLS estimates. Let $\nu_2 = u + \varepsilon_2$. Then, by construction, $\widehat{y}_2 = y_2 - \widehat{\nu}_2$. By replacing $\widehat{y}_2$ in (10.3), we get

$$y_1 = \beta_0 + y_2\beta_1 - \widehat{\nu}_2 + x\beta_2 + u + \varepsilon_1$$

Estimating this equation by OLS (treating $\widehat{\nu}$ as an additional control regressor) yields an estimate of β_1 that is identical to the 2SLS estimate. This estimator, the 2SRI, is a specific kind of control function estimator. We introduce 2SRI not merely to reproduce the 2SLS results but as a bridge to the discussion of endogeneity in models in which the endogenous regressor is better modeled nonlinearly, or more generally nonlinear models with endogenous regressors.

Further intuition can be found if we tease apart the error from the main equation into two pieces—the part correlated with the endogenous variable and the part that is independent. A control function is a variable (or variables) that approximates the correlated part. Newey, Powell, and Vella (1999) proved that there exists an optimal control function. If a researcher could observe such a variable, then including it in the main equation would be like including the omitted variable that caused the endogeneity. The remaining error would be uncorrelated with the endogenous variable.

We demonstrate the use of 2SRI in linear models using the example data. First, we estimate the first-stage regression of `y2` as a function of exogenous `x` and instrument `w`. We then compute residuals `nu2_hat`. We then estimate the outcome variable `y1` as a function of endogenous `y2`, exogenous `x`, and the estimated residual `nu2_hat`. As the results are very similar but not numerically identical to the 2SLS results. The estimated coefficient is close to 1.0, the true value. The standard error is also similar to the standard error found by 2SLS.

```
. *** ERM with FIML
. eregress y1 x, endogenous(y2 = x w)

Iteration 0:    log likelihood = -33888.338
Iteration 1:    log likelihood = -33888.338

Extended linear regression                       Number of obs    =      10,000
                                                 Wald chi2(2)     =   19172.17
Log likelihood = -33888.338                      Prob > chi2      =     0.0000
```

	Coef.	Std. Err.	z	P>|z|	[95% Conf.	Interval]
y1						
x	1.073925	.0765736	14.02	0.000	.9238432	1.224006
y2	.9323177	.0751241	12.41	0.000	.7850771	1.079558
_cons	-.0010483	.014614	-0.07	0.943	-.0296912	.0275946
y2						
x	1.002408	.0140734	71.23	0.000	.974825	1.029992
w	.194396	.0141223	13.77	0.000	.1667168	.2220753
_cons	-.0096731	.0141216	-0.68	0.493	-.037351	.0180048
var(e.y1)	2.132249	.1701171			1.823587	2.493155
var(e.y2)	1.993955	.0281988			1.939445	2.049997
corr(e.y2,						
e.y1)	.5404179	.0519153	10.41	0.000	.4309132	.6342377

10.3 Endogeneity with a binary endogenous variable

We next show how to fit a model with a continuous main dependent variable and a binary (nonlinear) endogenous variable. This example builds on several ideas presented in section 10.2. The outcome of interest, y_1, is continuous and determined by a linear equation as before,

$$y_1 = \beta_0 + y_2\beta_1 + x\beta_2 + u + \varepsilon_1$$

except that the covariate y_2 is now binary and takes only two values, 0 and 1. Let

$$y_2 = \mathbf{1}(\gamma_0 + w\gamma_1 + x\gamma_2 + u + \varepsilon_2) \tag{10.4}$$

where $\mathbf{1}(.)$ denotes the indicator function returning values of 1 if the argument is greater than 0, and 0 otherwise. Let the distribution of $u + \varepsilon_2$ be standard normal. In other words, y_2 is determined by a probit model. Even if ε_1 and ε_2 are uncorrelated with each other, the composite error terms, $u + \varepsilon_1$ and $u + \varepsilon_2$, are correlated, thus rendering OLS estimates of the outcome equation inconsistent.

We estimate the parameters of this model using three methods that account for the endogeneity of y_2. 2SLS in this context ignores the discrete nature of y_2. 2SRI and **eregress** with the **probit** option both explicitly model y_2 as a binary outcome. All three estimators are consistent, so our example provides only a sense of finite sample differences.

As in section 10.2.1, we use a simulated dataset with 10,000 observations. As the list of commands shown below demonstrates, we begin by drawing data for the observed covariates, x and w, from standard normal distributions. The unobserved variables, u, e1, and e2, are drawn from independent normal distributions with zero mean and variances equal to 0.5. Thus the variance of the sum of u and e2 equals 1, which is convenient for the interpretation of probit coefficients. We have also scaled the distribution of e1, but that has no substantive significance. The binary endogenous covariate y2, following (10.4), is a function of x, u, the unobserved error e2, and a covariate w. Finally, we generate the dependent variable, y1, using (10.1) as a function of an exogenous covariate x, the endogenous binary covariate y2, the unobserved covariate u, and the unobserved error e1.

```
. clear
. set obs 10000
number of observations (_N) was 0, now 10,000
. set seed 123456
. generate x  = rnormal(0,1)              // exogenous variable
. generate w  = rnormal(0,1)             // instrumental variable
. generate u  = sqrt(0.5)*rnormal(0,1)    // omitted variable
. generate e1 = sqrt(0.5)*rnormal(0,1)    // outcome eq error
. generate e2 = sqrt(0.5)*rnormal(0,1)    // endogenous regressor eq error
. generate y2st = -1.8 + x + .2*w + u + e2 // endogenous regressor latent eq
. generate y2 = y2st>0                    // binary variable from latent
. generate y1 = y2 + x + u + e1           // outcome eq
```

We calculate the frequency distribution of y2 to show that it is equal to 1 about 10% of the time. Similar rates are quite common in empirical research. We will describe its relevance below.

```
. tabulate y2
```

y2	Freq.	Percent	Cum.
0	8,960	89.60	89.60
1	1,040	10.40	100.00
Total	10,000	100.00	

We first estimate an OLS regression that does not account for the endogeneity of y2. We know that the estimates from such a regression will be inconsistent. We see that the OLS estimate of the coefficient on y2 is 1.76 and the confidence interval is far away from the true value of 1.0. Estimators that do not account for endogeneity can be misleading.

```
. *** ERM with FIML
. eregress y1 x, endogenous(y2 = x w, probit)

Iteration 0:    log likelihood = -16109.414
Iteration 1:    log likelihood = -16109.223
Iteration 2:    log likelihood = -16109.222

Extended linear regression                      Number of obs    =      10,000
                                                Wald chi2(2)     =    13163.22
Log likelihood = -16109.222                     Prob > chi2      =      0.0000
```

	Coef.	Std. Err.	z	P>|z|	[95% Conf.	Interval]
y1						
x	1.010564	.0134985	74.87	0.000	.9841071	1.03702
1.y2	.952913	.0692817	13.75	0.000	.8171233	1.088703
_cons	.0045567	.0123558	0.37	0.712	-.0196601	.0287736
y2						
x	1.062264	.0282402	37.62	0.000	1.006914	1.117613
w	.2148221	.0200246	10.73	0.000	.1755747	.2540696
_cons	-1.858288	.0316003	-58.81	0.000	-1.920223	-1.796352
var(e.y1)	1.000949	.0164973			.9691312	1.033811
corr(e.y2, e.y1)	.5022855	.0344776	14.57	0.000	.4316986	.5667582

We saw earlier that, while the 2SLS estimator is consistent, it does not produce precise estimates of the parameter of interest in this example. Both 2SRI and FIML estimators perform much better. However, the performance of the 2SLS estimator can be improved. Recall that y2 takes the value 1 about 10% of the time. It is in these ranges of rates of binary outcomes when nonlinear estimators like the probit have the most gains relative to an estimate from a linear probability model. But the predictive performance of the linear probability model can be improved by introducing nonlinear functions of the covariates. We consider a quadratic polynomial of x and w to illustrate. The first-stage results table shows the coefficient estimates from such a model. Two out of three of the higher-order terms are statistically significant. The results of the second-stage regression show that the point estimate of the coefficient on y2 is now 0.995—very close to 1.0. The standard error of the estimate is 0.28, which is a substantial improvement over the standard error in the first set of 2SLS.

```
. *** 2SLS
. ivregress 2sls y1 x c.x#c.x (y2 = w c.w#c.w c.w#c.x c.w#c.x#c.x), first
First-stage regressions
```

				Number of obs	=	10,000
				F(6, 9993)	=	643.51
				Prob > F	=	0.0000
				R-squared	=	0.2787
				Adj R-squared	=	0.2783
				Root MSE	=	0.2593

y2	Coef.	Std. Err.	t	P>|t|	[95% Conf. Interval]	
x	.1299141	.0025853	50.25	0.000	.1248464	.1349818
c.x#c.x	.0602623	.0018422	32.71	0.000	.0566512	.0638733
w	.0195407	.003206	6.10	0.000	.0132563	.0258251
c.w#c.w	.0011888	.0018143	0.66	0.512	-.0023676	.0047453
c.w#c.x	.0248459	.0025809	9.63	0.000	.0197868	.0299051
c.w#c.x#c.x	.0064948	.0018621	3.49	0.000	.0028447	.0101448
_cons	.0427943	.0036668	11.67	0.000	.0356067	.0499819

Instrumental variables (2SLS) regression				Number of obs	=	10,000
				Wald chi2(3)	=	13259.46
				Prob > chi2	=	0.0000
				R-squared	=	0.6034
				Root MSE	=	.99269

y1	Coef.	Std. Err.	z	P>|z|	[95% Conf. Interval]	
y2	1.090112	.269262	4.05	0.000	.5623678	1.617856
x	.9927859	.0363098	27.34	0.000	.92162	1.063952
c.x#c.x	-.0091724	.0177318	-0.52	0.605	-.0439262	.0255814
_cons	-.0005648	.0169706	-0.03	0.973	-.0338265	.032697

```
Instrumented: y2
Instruments:  x c.x#c.x w c.w#c.w c.w#c.x c.w#c.x#c.x
```

Readers should note a feature of the specification of the model using ivregress. The polynomial terms that involve w are included in the set of instruments that affect only y2 directly. The polynomial term that involves only x, c.x#c.x, appears in both the first and second stages of the regression because it is a common exogenous regressor. The estimated coefficient on c.x#c.x is very close to zero and not statistically significant in the regression of y1, as it should be. Including it in the fitted model, and not the true model, has no deleterious effect.

```
. *** GMM control function estimator with consistent standard errors
. gmm (r1: y2 - {xb1:w x _cons})
> (r2: (y1 - {b0} - {b1}*y2 - {b2}*x - {b3}*(y2-{xb1:}))))
> (r3: (y1 - {b0} - {b1}*y2 - {b2}*x - {b3}*(y2-{xb1:}))*
> (y2-{xb1:})),
> instruments(r1: w x)
> instruments(r2: y2 x )
> instruments(r3: )
> onestep winitial(identity)

Step 1
Iteration 0:    GMM criterion Q(b) = 55.751278  (not concave)
Iteration 1:    GMM criterion Q(b) = .28283334
Iteration 2:    GMM criterion Q(b) = .03763702
Iteration 3:    GMM criterion Q(b) = .00491432
Iteration 4:    GMM criterion Q(b) = 8.956e-26
Iteration 5:    GMM criterion Q(b) = 1.311e-31

note: model is exactly identified

GMM estimation

Number of parameters =    7
Number of moments    =    7
Initial weight matrix: Identity                  Number of obs   =     10,000
```

	Coef.	Robust Std. Err.	z	P>|z|	[95% Conf. Interval]	
w	.194396	.0143211	13.57	0.000	.1663272	.2224648
x	1.002408	.0139902	71.65	0.000	.9749882	1.029829
_cons	-.0096731	.0141233	-0.68	0.493	-.0373543	.0180081
/b0	-.0010483	.0146135	-0.07	0.943	-.0296903	.0275937
/b1	.9323174	.0747638	12.47	0.000	.7857831	1.078852
/b2	1.073925	.0762483	14.08	0.000	.924481	1.223369
/b3	.5588449	.075359	7.42	0.000	.411144	.7065458

```
Instruments for equation r1: w x _cons
Instruments for equation r2: y2 x _cons
Instruments for equation r3: _cons
```

In addition to fitting 2SRI for models with endogeneity, GMM can estimate multiple equations simultaneously. This can be useful in health economics for fitting two-part models all in one command. The Poisson count model is especially easy to fit with GMM; therefore, Poisson with IVs is also straightforward. However, other count models are not as easy to implement in GMM.

With GMM, there is an additional test statistic to test the overidentification assumption. Hansen's J statistic is a weighted average of the score function of the instruments times the residuals. Under homoskedasticity, the J statistic has the interpretation of the explained sum of squares from a regression of the 2SLS residuals on a vector of the instruments.

10.5 Stata resources

One Stata command to fit linear models with IVs is `ivregress`—which can be estimated with 2SLS, limited information maximum likelihood, or GMM. The `estat` commands, part of the `ivregress` postestimation commands, make it easy to run statistical tests of the main assumptions of strength and validity of the instruments. In addition, Stata has a unified set of commands, called ERM, that allow for estimation of linear and some nonlinear models, where the covariates can be exogenous or endogenous. The basic command for linear models is `eregress`, and the command that estimates treatment effects is `etregress`.

For nonlinear models, Stata will estimate 2SRI for probit models with the `ivprobit` command and the `twostep` option, for tobit models with the `ivtobit` command and the `twostep` option, and for Poisson models with the `ivpoisson` command and the `cfunction` option. The ERM commands can be used for probit and ordered probit models with endogenous covariates. The basic command for probit models is `eprobit`. The results from `ivprobit` without the `twostep` option is identical to `eprobit`, although the syntax is slightly different.

Stata has extensive capabilities of fitting GMM models, not just the GMM version of linear IVs. The `gmm` command (as opposed to the `gmm` option with `ivregress`) can estimate multiple equations simultaneously.

11 Design effects

11.1 Introduction

So far, we have described econometric methods and analyzed data from the 2004 Medical Expenditure Panel Survey (MEPS) dataset as if the data were collected using simple random sampling. However, in many research studies, either the design of the data or the study objectives may require the analyst to pay attention to features related to complex sampling. Specifically, observations in surveys are often not drawn with equal probability. For example, observations within households, hospitals, counties, and healthcare markets are correlated in unobserved ways—or the dataset is incomplete because of refusal, attrition, or item nonresponse. Such issues are often intrinsic in the design of survey data.

The literature on survey design issues and statistics for data from complex surveys is large and detailed (for example, Skinner, Holt, and Smith [1989]). On the other hand, the discussion of these issues in standard econometrics textbooks is sparse; exceptions include Cameron and Trivedi (2005), Deaton (1997), and Wooldridge (2010). Our objective here is not to survey that entire field of literature but rather to provide an introduction to the issues, intuition about the consequences of ignoring design effects, and some basic approaches to control for design effects through examples.

When large, multipurpose surveys such as the MEPS are conducted, a simple random sample is rarely collected for a variety of well-known financial and statistical efficiency reasons. When more complex sampling methods are used, estimates of model parameters, marginal and incremental effects, and inference may be more reliable if the specifics of the sampling design are accounted for during estimation. Indeed, if you ignore the sampling design entirely, the standard errors of parameter estimates will likely be underestimated, possibly leading to results that seem to be statistically significant, when in fact, they are not. The difference in point estimates and standard errors obtained using methods with and without survey-design features will vary from dataset to dataset and between variables within the same dataset. There is no practical way to know beforehand how different the results might be. Arguably, this lack of specificity has led to economists generally underemphasizing complex survey-design issues in econometrics. Solon, Haider, and Wooldridge (2015) have a recent discussion of these issues. Cameron and Trivedi (2005) note, however, that the effect of weighting tends to be much smaller in the regression context where the focus is on the relationship between a covariate and an outcome.

For many research questions and study designs, the correlations between groups of observations on unobserved attributes can play an important role in correct inference. Consider, for example, a typical survey of individuals in which households are randomly sampled, but all individuals within a household are surveyed. Suppose further that income is measured only at the household level but that health insurance status and healthcare use are both measured at the individual level (so there is variation within households). Then, income is, by definition, perfectly correlated within a household for every household. Thus, if a researcher is interested in the difference in income between insured and uninsured individuals or the income elasticity of healthcare use— where healthcare use is measured at the individual level—then not accounting for the correlation between individuals in a household will cause the standard errors of the estimates to be understated (Moulton 1986, 1990).

Furthermore, in many studies in which difference-in-differences designs (comparing treatment and control groups across preperiods and postperiods) are used, both the treatment and control groups may consist of sets of units that are considerably more aggregated than the unit of observation. For example, states or counties may be the level at which the treatment of interest is applied, but observations may be at the individual level. Thus the treatment assignment will be common to all units within the states or counties, causing observations within those units to be correlated. In such situations, too, summary statistics and estimates from regressions based on individual observations will yield standard errors that are too small, thus distorting the size of inference test statistics (Bertrand, Duflo, and Mullainathan 2004).

Below, we begin by describing some key features of many survey designs and the consequences of ignoring such design features. Then, we provide a number of examples using our MEPS sample—first in the context of calculating summary statistics and then in the regression context. Also note that some of these issues also occur in randomized controlled trial data, administratively collected data such as medical claims, and in data collected in other ways. Thus, although the descriptions below are framed in the survey data context, the issues apply to many empirical data analyses.

11.2   Features of sampling designs

Sampling designs for large surveys can be quite complex, but most of them share two features. First, observations are not sampled with equal probability; thus each observation is associated with a weight that indicates its relative importance in the sample relative to the population. Second, observations are not all sampled independently but instead are sampled in clusters. Stratification by subgroup is one important kind of clustering. Then, each observation in the sample is associated with a cluster identifier. Ignoring both weights and clusters can lead to misleading statistics and inference based on simple random sampling.

11.2.1   Weights

There are many types of weights that can be associated with a survey. Perhaps the most common is the sampling weight, which is used to denote the inverse of the probability of being included in the sample because of the sampling design. Therefore, observations that are oversampled will have lower weights than observations that are undersampled. In addition, postsampling adjustments to the weights are often made to adjust for deviations of the data-collection scheme from the original design. In Stata, `pweights` are sampling weights. Commands that allow `pweights` typically provide a `vce(cluster clustvar)` option, described below. Under many sampling designs, the sum of the sampling weights will equal the relevant population total instead of the sample size.

In MEPS, minorities were oversampled by design, so there were sufficient observations for each minority group to allow analysts to obtain reliable estimates of interest for each of those groups. Households containing Hispanics and blacks were oversampled at rates of approximately 2 and 1.5 times, respectively, the rate of remaining households. However, as a consequence, averages or counts taken over the entire sample without adjustments for the oversampling will not be generally representative of the population. Once sampling weights are accounted for the relatively large number of observations for minorities will be appropriately downweighted when aggregates and averages are estimated.

Many Stata commands also allow one or more of three additional types of weights: `fweights`, `aweights`, and `iweights`. We briefly describe the application of each below, but note that they are not generally considered as arising from complex survey methodology. Frequency weights (`fweights`) are integers representing the number of observations each sampled observation really represents. Analytic weights (`aweights`) are typically appropriate when each observation in the data is a summary statistic, such as the count or average, over a group of observations or to address issues of heteroskedasticity. The prototypical example is the instance of rates. For example, consider a county-level dataset in which each observation consists of rates that measure socioeconomic characteristics of people in the county in a particular year. Then, the weighting variable contains the number of individuals over which the average was calculated. Finally, most Stata commands allow the user to specify an importance weight (`iweight`). The `iweight` has no formal statistical definition but is assumed to reflect the importance of each observation in a sample.

11.2.2   Clusters and stratification

As mentioned above, individuals are not sampled independently in most survey designs. Collections of observational units (for example, states, counties, or households) are typically sampled as a group known as a cluster. The purpose of clustering is usually to save time and money in the sample collection, because it is often easier to collect information from people who live in close proximity to each other than to have a true random sample. There may also be further subsampling within the clusters. The clusters at the first level of sampling are called primary sampling units (PSUs). For example, in

11.3.1 Point estimation

To provide intuition, we present three examples of how weights affect point estimation. We start with the simplest case of using sampling weights to estimate a population average. Suppose that there are sampling weights, w_i, for observations $i = 1, 2, \ldots, N$. Note that the sample size is N, but let the total population be W, equal to the sum of all the individual weights:

$$W = \sum_{i=1}^{N} w_i$$

The population mean, $\bar{y}$, of a random variable, y_i, is the weighted average.

$$\bar{y} = \frac{1}{W} \sum_{i=1}^{N} w_i y_i$$

The *Stata User's Guide* has the estimators for additional, progressively more complex sampling designs.

The second example is for weighted least squares. The least-squares estimator for linear regression can also be readily modified to incorporate sampling weights. In Stata, the observations denoting weights are normalized to sum to N, if `pweights` or `aweights` are specified. Let w_i denote the normalized or unnormalized weights, let $\mathbf{w}$ denote the vector of weights, let $\mathbf{D} = \text{diag}(\mathbf{w})$, and let $\mathbf{X}$ be the matrix of covariates, $\mathbf{x}$. The goal is to estimate the vector of parameters, $\boldsymbol{\beta}$, in the linear model, $y_i = \mathbf{x}_i' \boldsymbol{\beta} + \epsilon_i$. Then, the estimated weighted least-squares parameters are found by the following formula.

$$\widehat{\boldsymbol{\beta}} = (\mathbf{X'DX})^{-1}(\mathbf{X'D}y)$$

Finally, consider the logistic regression as an example of a maximum likelihood estimator. The log- (pseudo-) likelihood function for the logistic distribution, F, with sampling weights is

$$\ln L = \sum_{i=1}^{N} w_i \ln F \left\{ (2y_i - 1)\, \mathbf{x}_i' \boldsymbol{\beta} \right\}$$

where y_i is a $(0, 1)$ indicator for the dependent variable, $z_i = \{(2y_i - 1)\mathbf{x}_i'\boldsymbol{\beta}\}$, and

$$F(z_i) = \frac{e^{z_i}}{1 + e^{z_i}}$$

In all three of these examples, weights (but not clustering) affect the point estimates. As we will demonstrate in section 11.4, Stata makes it easy to incorporate weights into all of these estimators.

11.3.2 Standard errors

Adjusting for weights and for clustering changes the standard errors of the estimates. The most commonly applied method to obtain the covariance matrix of $\widehat{\boldsymbol{\beta}}$ involves a Taylor-series based linearization (popularly known as the delta method), in which weights and clustering are easily incorporated. This is implemented as a standard option in Stata. After declaring the survey design with the `svyset` command, use the `vce(cluster clustvar)` option.

When the parameters of interest are complex functions of the model parameters, the linearized variance estimation may not be convenient. In such situations, bootstrap or jackknife variance estimation can be important nonparametric ways to obtain standard errors. With few assumptions, bootstrap and jackknife resampling techniques provide a way of estimating standard errors and other measures of statistical precision and are especially convenient when no standard formula is available. We begin by describing the bootstrap, which is described in detail in Cameron and Trivedi (2005).

The principle of bootstrap resampling is quite simple. Consider a dataset with N observations, on which a statistical model is fit and parameters (or functions of parameters such as marginal effects) are calculated. The bootstrap procedure involves drawing, with replacement—so that the same observation may be drawn again—N observations from the N-observation dataset. The model and parameters of interest are estimated from this resampled dataset. This process is repeated many times, and the empirical distribution of the parameters of interest is used to calculate standard deviations or other features of the distributions of the parameters.

The jackknife method is, like the bootstrap, a technique that is independent of the estimation procedure. In it, the model is fit multiple times, with one observation being dropped from the estimation sample each time. The standard errors of the estimates of interest are then calculated as the empirical standard deviations of the estimates over the set of replicates.

Both the bootstrap and jackknife methods are easily adjusted to deal with clustering. In the context of survey designs with clustering, the unit of observation for resampling in the bootstrap and jackknife is a cluster or a PSU. If the survey design also involves sampling weights, both the bootstrap and jackknife methods become considerably more complex to implement. For each replication, the sampling weights need to be adjusted, because some clusters may be repeated, while others may not be in the sample in the case of the bootstrap, or because one cluster is dropped from the replicate sample in the case of the jackknife. Some complex surveys provide bootstrap or jackknife replicate weights, in which case those methods can be implemented in the complex survey context using `svy bootstrap` or `svy jackknife`. If the resampled weights are not provided, the researcher must calculate those weights. This requires in-depth knowledge of the survey design and the way in which the weights were originally constructed.

Although we do not show an empirical example of either bootstrap or jackknife methods in this chapter, the Stata manual has good examples in the *Stata Base Reference Manual* under `bootstrap` and `jackknife`.

```
. estimates store weights
. quietly mean exp_tot race_bl_pct [pw=wtdper], vce(cluster clusterid)
. estimates store clust_wgt
. quietly svy: mean exp_tot race_bl_pct
. estimates store survey
. estimates table *, b(%7.1f) modelwidth(9)
>            title(Alternative cluster and weight options: Sample means)
Alternative cluster and weight options: Sample means
```

Variable	noadjust	cluster	weights	clust_wgt	survey
exp_tot	3685.2	3685.2	3838.9	3838.9	3838.9
race_bl_pct	13.8	13.8	10.8	10.8	10.8

In our next example, we estimate mean total expenditures by race to conduct a t test for whether the difference in means is different from zero. We find that mean spending for nonblacks is \$3,731 and mean spending for blacks is \$3,402. The difference in spending is not statistically significant at the traditional 5% level, but it is significant at the 10% level.

```
. *** Sample mean and simple t tests
. mean exp_tot, over(race_bl)
Mean estimation                        Number of obs   =      19,386

     _subpop_1: race_bl = Not black race
     _subpop_2: race_bl = Black race
```

Over	Mean	Std. Err.	[95% Conf. Interval]	
exp_tot				
_subpop_1	3730.723	77.56493	3578.689	3882.757
_subpop_2	3401.788	154.1159	3099.707	3703.868

```
. test [exp_tot]_subpop_1 = [exp_tot]_subpop_2
 ( 1)  [exp_tot]_subpop_1 - [exp_tot]_subpop_2 = 0

       F(  1, 19385) =     3.63
            Prob > F =    0.0566
```

We repeat the analysis taking sampling weights and clustering into account using the svy prefix. The results are quite remarkable. The weighted sample mean for the subsample of blacks (race_bl=1) is smaller than the unweighted sample mean (3,115 compared with 3,402), while the weighted sample mean for nonblacks is larger than its unweighted counterpart (3,927 compared with 3,731). Thus, even though the cluster-corrected standard errors of the means estimated using the survey-aware methods are uniformly larger than the naïve estimates, the t statistic for the difference in expenditures between blacks and nonblacks is not statistically significant at 5% when the naïve estimates are used, but is significant at the 0.1% level when survey-aware estimates are used.

```
. *** Sample mean and simple t tests incorporating survey features
. svy: mean exp_tot, over(race_bl)
(running mean on estimation sample)

Survey: Mean estimation

Number of strata =      203      Number of obs   =       19,386
Number of PSUs   =      448      Population size = 187,973,715
                                 Design df       =          245

     _subpop_1: race_bl = Not black race
     _subpop_2: race_bl = Black race
```

Over	Mean	Linearized Std. Err.	[95% Conf. Interval]	
exp_tot				
_subpop_1	3926.681	110.3909	3709.244	4144.117
_subpop_2	3114.646	168.0148	2783.708	3445.583

```
. test [exp_tot]_subpop_1 = [exp_tot]_subpop_2

Adjusted Wald test

 ( 1)  [exp_tot]_subpop_1 - [exp_tot]_subpop_2 = 0

       F(  1,    245) =    15.87
             Prob > F =     0.0001
```

11.4.3   Weighted least-squares regression

Common wisdom is that design effects matter less in regression contexts than when summary statistics are desired (Solon, Haider, and Wooldridge 2015; Cameron and Trivedi 2005), but—aside from limited special cases—there is no formal derivation of this understanding, nor can the differences between naïve and more sophisticated estimates be signed a priori. Therefore, it may be better to take design effects seriously, regardless of the nature of the statistical model under consideration. Below, we demonstrate the effects of taking survey design features into consideration. We show this first in the context of a linear regression in which the coefficient estimates themselves are of primary interest and show this second in the context of a Poisson regression—in which incremental and marginal effects are of interest.

We estimate regressions of total expenditures by self or family (exp_self) on continuous age (age) and indicators for gender (female), black race (race_bl), and South region (reg_south) using ordinary and weighted least squares and using alternative formula to estimate standard errors. To facilitate comparison of the estimates from the different methods, we accumulate regression results into a table. The first regression (robust) does not account for any design features but does estimate robust standard errors. The second set of estimates (cluster) are obtained using ordinary least squares, but the standard errors take clustering into account. The third set of estimates (weights) are calculated using weighted least squares; the weights are probability or sampling weights. These estimates do not account for clustering. The fourth set of estimates (clust_wgt) incorporate both cluster–robust standard errors and sampling weights as options of regress but without the svy: prefix. The final specification (survey) uses the svy:

prefix to produce fully survey-aware estimates that control for weights, clustering, and stratification.

The results show that, as expected, adjusting for weights changes the point estimates, while adjusting for clustering and stratification does not. The point estimates are the same in columns 1 and 2 (no weights) but different from the point estimates in columns 3–5 (with weights).

In contrast, clustering changes only the standard errors. In fact, estimates of standard errors are also not substantially different for most variables. However, for reg_south, the standard errors increase in a substantial way. Recall that our clustering variable is designed to have geographic and race implications but should have no implications for age and gender for this MEPS sample. Therefore, it is not surprising that the standard errors on the coefficients on age and gender do not change much and that the standard error on the coefficient of reg_south increases. It is a bit surprising that the standard error of race_bl does not change much.

The landscape changes quite a bit once sampling weights are introduced. All point estimates and standard errors change substantially, except for those on age. The implications for interpretation of the effect of race_bl are substantial. While one would comfortably conclude that black race was not associated with healthcare expenditures from columns 1 and 2 (p-value of 0.58), one would have to think about the association further given the p-values less than or equal to 0.1 in the remaining three cases.

```
. *** Linear regression
. quietly regress exp_tot age i.female i.race_bl i.reg_south, vce(robust)
. estimates store robust
. quietly regress exp_tot age i.female i.race_bl i.reg_south,
>           vce(cluster clusterid)
. estimates store cluster
. quietly regress exp_tot age i.female i.race_bl i.reg_south [pw=wtdper]
. estimates store weights
. quietly regress exp_tot age i.female i.race_bl i.reg_south [pw=wtdper],
>           vce(cluster clusterid)
. estimates store clust_wgt
. quietly svy: regress exp_tot age i.female i.race_bl i.reg_south
. estimates store survey
```

```
. estimates table *, b(%7.2f) se(%7.2f) p(%7.4f) modelwidth(9) drop(_cons)
> title(Alternative cluster and weight options: Linear regression estimates)
Alternative cluster and weight options: Linear regression estimates
```

Variable	robust	cluster	weights	clust_wgt	survey
age	129.13	129.13	129.91	129.91	129.91
	4.68	4.52	5.62	5.33	5.25
	0.0000	0.0000	0.0000	0.0000	0.0000
female					
Female	895.51	895.51	769.09	769.09	769.09
	138.29	134.52	188.06	184.53	188.00
	0.0000	0.0000	0.0000	0.0000	0.0001
race_bl					
Black race	-92.19	-92.19	-317.78	-317.78	-317.78
	167.13	167.38	182.56	196.65	188.59
	0.5812	0.5820	0.0818	0.1068	0.0933
reg_south					
South	-256.71	-256.71	-325.40	-325.40	-325.40
	132.97	145.14	164.93	163.80	158.92
	0.0536	0.0776	0.0485	0.0476	0.0417

```
                                                          legend: b/se/p
```

11.4.4   Weighted Poisson count model

We estimated Poisson regressions of the counts of office-based provider visits (use_off) on female, race_bl, and reg_south with and without survey design features. In each case, we estimated sample average partial effects of age and female using margins, which naturally accounts for survey design features used to obtain the model parameter estimates if the regression is specified this way. The results, tabulated below, show that the marginal effects of age are not noticeably different across procedures; neither are the associated standard errors. The incremental effects of being female and their associated standard errors increase modestly once sampling weights are taken into account.

As with the previous linear regression example, the most dramatic differences across specifications are seen in the effects of black race and South region. The incremental effect of race_bl increases substantially in magnitude once sampling weights are introduced. Standard errors also increase but not by much. The incremental effects of reg_south do not change much at all across specifications, but their standard errors increase once clustering is accounted for.

References

Abrevaya, J. 2002. Computing marginal effects in the Box–Cox model. *Econometric Reviews* 21: 383–393.

Ai, C., and E. C. Norton. 2000. Standard errors for the retransformation problem with heteroscedasticity. *Journal of Health Economics* 19: 697–718.

———. 2003. Interaction terms in logit and probit models. *Economics Letters* 80: 123–129.

———. 2008. A semiparametric derivative estimator in log transformation models. *Econometrics Journal* 11: 538–553.

Akaike, H. 1970. Statistical predictor identification. *Annals of the Institute of Statistical Mathematics* 22: 203–217.

Angrist, J. D., and A. B. Krueger. 2001. Instrumental variables and the search for identification: From supply and demand to natural experiments. *Journal of Economic Perspectives* 15(4): 69–85.

Angrist, J. D., and J.-S. Pischke. 2009. *Mostly Harmless Econometrics: An Empiricist's Companion*. Princeton: Princeton University Press.

Arlot, S., and A. Celisse. 2010. A survey of cross-validation procedures for model selection. *Statistics Surveys* 4: 40–79.

Barnett, S. B. L., and T. A. Nurmagambetov. 2011. Costs of asthma in the United States: 2002–2007. *Journal of Allergy and Clinical Immunology* 127: 145–152.

Bassett, G., Jr., and R. Koenker. 1982. An empirical quantile function for linear models with iid errors. *Journal of the American Statistical Association* 77: 407–415.

Basu, A., and P. J. Rathouz. 2005. Estimating marginal and incremental effects on health outcomes using flexible link and variance function models. *Biostatistics* 6: 93–109.

Baum, C. F. 2006. *An Introduction to Modern Econometrics Using Stata*. College Station, TX: Stata Press.

Belotti, F., P. Deb, W. G. Manning, and E. C. Norton. 2015. twopm: Two-part models. *Stata Journal* 15: 3–20.

Berk, M. L., and A. C. Monheit. 2001. The concentration of health care expenditures, revisited. *Health Affairs* 20: 9–18.

Bertrand, M., E. Duflo, and S. Mullainathan. 2004. How much should we trust differences-in-differences estimates? *Quarterly Journal of Economics* 119: 249–275.

Bitler, M. P., J. B. Gelbach, and H. W. Hoynes. 2006. Welfare reform and children's living arrangements. *Journal of Human Resources* 41: 1–27.

Blough, D. K., C. W. Madden, and M. C. Hornbrook. 1999. Modeling risk using generalized linear models. *Journal of Health Economics* 18: 153–171.

Blundell, R. W., and R. J. Smith. 1989. Estimation in a class of simultaneous equation limited dependent variable models. *Review of Economic Studies* 56: 37–57.

―――. 1994. Coherency and estimation in simultaneous models with censored or qualitative dependent variables. *Journal of Econometrics* 64: 355–373.

Bound, J., D. A. Jaeger, and R. M. Baker. 1995. Problems with instrumental variables estimation when the correlation between the instruments and the endogenous explanatory variable is weak. *Journal of the American Statistical Association* 90: 443–450.

Box, G. E. P., and D. R. Cox. 1964. An analysis of transformations. *Journal of the Royal Statistical Society, Series B* 26: 211–252.

Box, G. E. P., and N. R. Draper. 1987. *Empirical Model-building and Response Surfaces.* New York: Wiley.

Buntin, M. B., and A. M. Zaslavsky. 2004. Too much ado about two-part models and transformation? Comparing methods of modeling Medicare expenditures. *Journal of Health Economics* 23: 525–542.

Cameron, A. C., J. B. Gelbach, and D. L. Miller. 2008. Bootstrap-based improvements for inference with clustered errors. *Review of Economics and Statistics* 90: 414–427.

―――. 2011. Robust inference with multiway clustering. *Journal of Business and Economic Statistics* 29: 238–249.

Cameron, A. C., and P. K. Trivedi. 2005. *Microeconometrics: Methods and Applications.* New York: Cambridge University Press.

―――. 2010. *Microeconometrics using Stata, Revised Edition.* Stata Press.

―――. 2013. *Regression Analysis of Count Data.* 2nd ed. Cambridge: Cambridge University Press.

Cattaneo, M. D., D. M. Drukker, and A. D. Holland. 2013. Estimation of multivalued treatment effects under conditional independence. *Stata Journal* 13: 407–450.

Cattaneo, M. D., and M. Jansson. 2017. Kernel-based semiparametric estimators: Small bandwidth asymptotics and bootstrap consistency. Working Paper. http://eml.berkeley.edu/~mjansson/Papers/ CattaneoJansson_BootstrappingSemiparametrics.pdf.

Cawley, J., and C. Meyerhoefer. 2012. The medical care costs of obesity: An instrumental variables approach. *Journal of Health Economics* 31: 219–230.

Claeskens, G., and N. L. Hjort. 2008. *Model Selection and Model Averaging*. Cambridge: Cambridge University Press.

Cole, J. A., and J. D. F. Sherriff. 1972. Some single- and multi-site models of rainfall within discrete time increments. *Journal of Hydrology* 17: 97–113.

Cook, P. J., and M. J. Moore. 1993. Drinking and schooling. *Journal of Health Economics* 12: 411–429.

Cox, N. J. 2004. Speaking Stata: Graphing model diagnostics. *Stata Journal* 4: 449–475.

Cragg, J. G. 1971. Some statistical models for limited dependent variables with application to the demand for durable goods. *Econometrica* 39: 829–844.

Dall, T. M., Y. Zhang, Y. J. Chen, W. W. Quick, W. G. Yang, and J. Fogli. 2010. The economic burden of diabetes. *Health Affairs* 29: 297–303.

Deaton, A. 1997. *The Analysis of Household Surveys: A Microeconometric Approach to Development Policy*. Washington, DC: The World Bank.

Deb, P. 2007. fmm: Stata module to estimate finite mixture models. Statistical Software Components S456895, Department of Economics, Boston College. https://ideas.repec.org/c/boc/bocode/s456895.html.

Deb, P., and P. K. Trivedi. 1997. Demand for medical care by the elderly: A finite mixture approach. *Journal of Applied Econometrics* 12: 313–336.

―――. 2002. The structure of demand for health care: latent class versus two-part models. *Journal of Health Economics* 21: 601–625.

Donald, S. G., D. A. Green, and H. J. Paarsch. 2000. Differences in wage distributions between Canada and the United States: An application of a flexible estimator of distribution functions in the presence of covariates. *Review of Economic Studies* 67: 609–633.

Dow, W. H., and E. C. Norton. 2003. Choosing between and interpreting the Heckit and two-part models for corner solutions. *Health Services and Outcomes Research Methodology* 4: 5–18.

Dowd, B. E., W. H. Greene, and E. C. Norton. 2014. Computation of standard errors. *Health Services Research* 49: 731–750.

Drukker, D. M. 2014. mqgamma: Stata module to estimate quantiles of potential-outcome distributions. Statistical Software Components S457854, Department of Economics, Boston College.
https://ideas.repec.org/c/boc/bocode/s457854.html.

———. 2016. Quantile regression allows covariate effects to differ by quantile. The Stata Blog: Not Elsewhere Classified. http://blog.stata.com/2016/09/27/quantile-regression-allows-covariate-effects-to-differ-by-quantile/.

———. 2017. Two-part models are robust to endogenous selection. *Economics Letters* 152: 71–72.

———. Forthcoming. Quantile treatment effect estimation from censored data by regression adjustment. *Stata Journal*.

Duan, N. 1983. Smearing estimate: A nonparametric retransformation method. *Journal of the American Statistical Association* 78: 605–610.

Duan, N., W. G. Manning, C. N. Morris, and J. P. Newhouse. 1984. Choosing between the sample-selection model and the multi-part model. *Journal of Business and Economic Statistics* 2: 283–289.

Efron, B. 1988. Logistic regression, survival analysis, and the Kaplan–Meier curve. *Journal of the American Statistical Association* 83: 414–425.

Enami, K., and J. Mullahy. 2009. Tobit at fifty: A brief history of Tobin's remarkable estimator, of related empirical methods, and of limited dependent variable econometrics in health economics. *Health Economics* 18: 619–628.

Ettner, S. L., G. Denmead, J. Dilonardo, H. Cao, and A. J. Belanger. 2003. The impact of managed care on the substance abuse treatment patterns and outcomes of Medicaid beneficiaries: Maryland's health choice program. *Journal of Behavioral Health Services and Research* 30: 41–62.

Ettner, S. L., R. G. Frank, T. G. McGuire, J. P. Newhouse, and E. H. Notman. 1998. Risk adjustment of mental health and substance abuse payments. *Inquiry* 35: 223–239.

Fan, J., and I. Gijbels. 1996. *Local Polynomial Modelling and Its Applications*. New York: Chapman & Hall/CRC.

Fenton, J. J., A. F. Jerant, K. D. Bertakis, and P. Franks. 2012. The cost of satisfaction: A national study of patient satisfaction, health care utilization, expenditures, and mortality. *Archives of Internal Medicine* 172: 405–411.

van Garderen, K. J., and C. Shah. 2002. Exact interpretation of dummy variables in semilogarithmic equations. *Econometrics Journal* 5: 149–159.

Garrido, M. M., P. Deb, J. F. Burgess, Jr., and J. D. Penrod. 2012. Choosing models for health care cost analyses: Issues of nonlinearity and endogeneity. *Health Services Research* 47: 2377–2397.

Gelman, A., and J. Hill. 2007. *Data Analysis Using Regression and Multilevel/Hierarchical Models*. Cambridge, UK: Cambridge University Press.

Gilleskie, D. B., and T. A. Mroz. 2004. A flexible approach for estimating the effects of covariates on health expenditures. *Journal of Health Economics* 23: 391–418.

Goldberger, A. S. 1981. Linear regression after selection. *Journal of Econometrics* 15: 357–366.

Gourieroux, C., A. Monfort, and A. Trognon. 1984a. Pseudo maximum likelihood methods: Applications to Poisson models. *Econometrica* 52: 701–720.

———. 1984b. Pseudo maximum likelihood methods: Theory. *Econometrica* 52: 681–700.

Greene, W. H. 2012. *Econometric Analysis*. 7th ed. Upper Saddle River, NJ: Prentice Hall.

Gurmu, S. 1997. Semi-parametric estimation of hurdle regression models with an application to Medicaid utilization. *Journal of Applied Econometrics* 12: 225–242.

Halvorsen, R., and R. Palmquist. 1980. The interpretation of dummy variables in semilogarithmic equations. *American Economic Review* 70: 474–475.

Hardin, J. W., and J. M. Hilbe. 2012. *Generalized Linear Models and Extensions*. 3rd ed. College Station, TX: Stata Press.

Hay, J. W., and R. J. Olsen. 1984. Let them eat cake: A note on comparing alternative models of the demand for medical care. *Journal of Business and Economic Statistics* 2: 279–282.

Heckman, J. J. 1979. Sample selection bias as a specification error. *Econometrica* 47: 153–161.

Heckman, J. J., and R. Robb, Jr. 1985. Alternative methods for evaluating the impact of interventions: An overview. *Journal of Econometrics* 30: 239–267.

Heckman, J. J., and E. J. Vytlacil. 2007. Econometric evaluation of social programs, part I: Causal models, structural models and econometric policy evaluation. In *Handbook of Econometrics, vol. 6B*, ed. J. J. Heckman and E. Leamer, 4779–4874. Amsterdam: Elsevier.

Hoch, J. S., A. H. Briggs, and A. R. Willan. 2002. Something old, something new, something borrowed, something blue: A framework for the marriage of health econometrics and cost-effectiveness analysis. *Health Economics* 11: 415–430.

Miranda, A. 2006. qcount: Stata program to fit quantile regression models for count data. *Statistical Software Components* S456714, Department of Economics, Boston College. https://ideas.repec.org/c/boc/bocode/s456714.html.

Moulton, B. R. 1986. Random group effects and the precision of regression estimates. *Journal of Econometrics* 32: 385–397.

———. 1990. An illustration of a pitfall in estimating the effects of aggregate variables on micro units. *Review of Economics and Statistics* 72: 334–338.

Mroz, T. A. 2012. A simple, flexible estimator for count and other ordered discrete data. *Journal of Applied Econometrics* 27: 646–665.

Mullahy, J. 1997. Heterogeneity, excess zeros, and the structure of count data models. *Journal of Applied Econometrics* 12: 337–350.

———. 1998. Much ado about two: Reconsidering retransformation and the two-part model in health econometrics. *Journal of Health Economics* 17: 247–281.

———. 2015. In memoriam: Willard G. Manning, 1946–2014. *Health Economics* 24: 253–257.

Murray, M. P. 2006. Avoiding invalid instruments and coping with weak instruments. *Journal of Economic Perspectives* 20: 111–132.

Nelson, C. R., and R. Startz. 1990. Some further results on the exact small sample properties of the instrumental variable estimator. *Econometrica* 58: 967–976.

Newey, W. K., J. L. Powell, and F. Vella. 1999. Nonparametric estimation of triangular simultaneous equations models. *Econometrica* 67: 565–603.

Newhouse, J. P., and M. McClellan. 1998. Econometrics in outcomes research: The use of instrumental variables. *Annual Review of Public Health* 19: 17–34.

Newhouse, J. P., and C. E. Phelps. 1976. New estimates of price and income elasticities of medical care services. In *The Role of Health Insurance in the Health Services Sector*, ed. R. N. Rosett, 261–320. Cambridge, MA: National Bureau of Economic Research.

Newson, R. 2003. Confidence intervals and p-values for delivery to the end user. *Stata Journal* 3: 245–269.

Norton, E. C., H. Wang, and C. Ai. 2004. Computing interaction effects and standard errors in logit and probit models. *Stata Journal* 4: 154–167.

Park, R. E. 1966. Estimation with heteroscedastic error terms. *Econometrica* 34: 888.

Picard, R. R., and R. D. Cook. 1984. Cross-validation of regression models. *Journal of the American Statistical Association* 79: 575–583.

Pohlmeier, W., and V. Ulrich. 1995. An econometric model of the two-part decision-making process in the demand for health care. *Journal of Human Resources* 30: 339–361.

Poirier, D. J., and P. A. Ruud. 1981. On the appropriateness of endogenous switching. *Journal of Econometrics* 16: 249–256.

Pregibon, D. 1981. Logistic regression diagnostics. *Annals of Statistics* 9: 705–724.

Racine, J., and Q. Li. 2004. Nonparametric estimation of regression functions with both categorical and continuous data. *Journal of Econometrics* 119: 99–130.

Ramsey, J. B. 1969. Tests for specification errors in classical linear least-squares regression analysis. *Journal of the Royal Statistical Society, Series B* 31: 350–371.

Rao, C. R., and Y. Wu. 2001. On model selection. *Lecture Notes-Monograph Series* 38: 1–64.

Roy, A., P. Sheffield, K. Wong, and L. Trasande. 2011. The effects of outdoor air pollutants on the costs of pediatric asthma hospitalizations in the united states, 1999–2007. *Medical Care* 49: 810–817.

Rubin, D. B. 1974. Estimating causal effects of treatments in randomized and nonrandomized studies. *Journal of Educational Psychology* 66: 688–701.

Schwarz, G. 1978. Estimating the dimension of a model. *Annals of Statistics* 6: 461–464.

Sin, C.-Y., and H. White. 1996. Information criteria for selecting possibly misspecified parametric models. *Journal of Econometrics* 71: 207–225.

Skinner, C. J., D. Holt, and T. M. F. Smith. 1989. *Analysis of Complex Surveys.* New York: Wiley.

Solon, G., S. J. Haider, and J. M. Wooldridge. 2015. What are we weighting for? *Journal of Human Resources* 50: 301–316.

Staiger, D. O., and J. H. Stock. 1997. Instrumental variables regression with weak instruments. *Econometrica* 65: 557–586.

Stock, J. H., J. H. Wright, and M. Yogo. 2002. A survey of weak instruments and weak identification in generalized method of moments. *Journal of Business and Economic Statistics* 20: 518–529.

Teicher, H. 1963. Identifiability of finite mixtures. *Annals of Mathematical Statistics* 34: 1265–1269.

Terza, J. V., A. Basu, and P. J. Rathouz. 2008. Two-stage residual inclusion estimation: Addressing endogeneity in health econometric modeling. *Journal of Health Economics* 27: 531–543.

Subject index